# About the Author

 Mr. Harte is the president of Althos, an expert information provider which researches, trains, and publishes on technology and business industries. He has over 29 years of technology analysis, development, implementation, and business management experience. Mr. Harte has worked for leading companies including Ericsson/General Electric, Audiovox/Toshiba and Westinghouse and has consulted for hundreds of other companies. Mr. Harte continually researches, analyzes, and tests new communication technologies, applications, and services. He has authored over 80 books on telecommunications technologies and business systems covering topics such as mobile telephone systems, data communications, voice over data networks, broadband, prepaid services, billing systems, sales, and Internet marketing. Mr. Harte holds many degrees and certificates including an Executive MBA from Wake Forest University (1995) and a BSET from the University of the State of New York, (1990).

# IPTV Basics

## By Lawrence Harte

Althos Publishing
Fuquay-Varina, NC 27526 USA
Telephone: 1-800-227-9681
Fax: 1-919-557-2261
email: info@althos.com
web: www.Althos.com

**Althos**

Copyright © 2006, 2007 By Althos Publishing
First Printing

Printed and Bound by Lightning Source, TN.

Every effort has been made to make this manual as complete and as accurate as possible. However, there may be mistakes both typographical and in content. Therefore, this text should be used only as a general guide and not as the ultimate source of IP Television information. Furthermore, this manual contains information on telecommunications accurate only up to the printing date. The purpose of this manual to educate. The authors and Althos Publishing shall have neither liability nor responsibility to any person or entity with respect to any loss or damage caused, or alleged to be caused, directly or indirectly by the information contained in this book.

**International Standard Book Number: 1-932813-56-X**

# Table of Contents

**INTERNET PROTOCOL TELEVISION(IPTV)** . . . . . . . . . . . . . 1

WHY CONSIDER IPTV AND INTERNET TELEVISION SERVICES . . . . . . . 6

*More Channels* . . . . . . . . . . . . . . . . . . . . . . . . . .7

*More Control* . . . . . . . . . . . . . . . . . . . . . . . . . .7

*More Services* . . . . . . . . . . . . . . . . . . . . . . . . . .9

HOW IPTV AND INTERNET TELEVISION SYSTEMS WORK . . . . . . . . 10

*Digitization - Converting Video Signals and Audio Signals to Digital Signals* . . . . . . . . . . . . . . . . . . . . . . . .11

*Digital Media Compression – Gaining Efficiency* . . . . . . . . .11

*Sending Packets* . . . . . . . . . . . . . . . . . . . . . . .13

*Converting Packets to Television Service* . . . . . . . . . . . .17

*Managing the Television Connections* . . . . . . . . . . . . . .18

*Multiple IPTVs per Home* . . . . . . . . . . . . . . . . . . .22

*Transmission* . . . . . . . . . . . . . . . . . . . . . . . .23

VIEWING IPTV . . . . . . . . . . . . . . . . . . . . . . . . . 26

*Multimedia Computer* . . . . . . . . . . . . . . . . . . . . .27

*IP Set Top Boxes (IP STB)* . . . . . . . . . . . . . . . . . . .28

*IP Televisions* . . . . . . . . . . . . . . . . . . . . . . . .28

*Mobile Video Telephones* . . . . . . . . . . . . . . . . . . .28

## IPTV SERVICES AND FEATURES ...................... 31

Television Programming ......................... 31
Enterprise (Company) Television ................ 32
Gaming ........................................ 32
Security Monitoring ........................... 32
Advertising ................................... 34
Television Commerce (T-Commerce) .............. 35
IPTV CONTENT .................................... 37
Network Programming ........................... 37
Syndicated Programming ........................ 37
Local Programming ............................. 37
Sponsored Programming ......................... 38
International Programming ...................... 38
Community Content ............................. 38
Personal Media ................................ 38
Long Tail Content ............................. 39
Flat Tail Content ............................. 39
IPTV SYSTEM FEATURES ............................ 39
Instant Activation ............................ 40
Real Time Accounting and Billing .............. 41
Channel Selection (Program Guides) ............ 42
Interactive Television (iTV) .................. 43
Anywhere Television Service (TV Roaming) ...... 43
Global Television Channels .................... 45
Personal Media Channels (PMC) ................. 46
Addressable Advertising ....................... 47
Ad Telescoping ................................ 49
Everything on Demand (EoD) .................... 49
Ad Bidding .................................... 51
T-Commerce Order Processing ................... 53
User Profiling ................................ 57

## INTRODUCTION TO IP VIDEO . . . . . . . . . . . . . . . . . . . . . . . 59

   *Resolution* . . . . . . . . . . . . . . . . . . . . . . . . . . . . . . . . . . . . . *60*
   *Frame Rates* . . . . . . . . . . . . . . . . . . . . . . . . . . . . . . . . . . . *60*
   *Aspect Ratio* . . . . . . . . . . . . . . . . . . . . . . . . . . . . . . . . . . . *62*
   *Letterbox* . . . . . . . . . . . . . . . . . . . . . . . . . . . . . . . . . . . . . . *63*
  ANALOG VIDEO . . . . . . . . . . . . . . . . . . . . . . . . . . . . . . . . . 64
   *Component Video* . . . . . . . . . . . . . . . . . . . . . . . . . . . . . . . *65*
   *Composite Video* . . . . . . . . . . . . . . . . . . . . . . . . . . . . . . . . *65*
   *Separate Video (S-Video)* . . . . . . . . . . . . . . . . . . . . . . . . *66*
   *Interlacing* . . . . . . . . . . . . . . . . . . . . . . . . . . . . . . . . . . . . . *66*
   *NTSC Video* . . . . . . . . . . . . . . . . . . . . . . . . . . . . . . . . . . . *67*
   *PAL Video* . . . . . . . . . . . . . . . . . . . . . . . . . . . . . . . . . . . . *68*
   *SECAM Video* . . . . . . . . . . . . . . . . . . . . . . . . . . . . . . . . . *69*
  DIGITAL VIDEO . . . . . . . . . . . . . . . . . . . . . . . . . . . . . . . . . . 69
   *4:2:2 Digital Video Format* . . . . . . . . . . . . . . . . . . . . . . . *69*
   *4:2:0 Digital Video Format* . . . . . . . . . . . . . . . . . . . . . . . *70*
   *Source Intermediate Format (SIF)* . . . . . . . . . . . . . . . . . *71*
   *Common Intermediate Format (CIF)* . . . . . . . . . . . . . . . *72*
   *Quarter Common Intermediate Format (QCIF)* . . . . . . . *73*
  VIDEO DIGITIZATION . . . . . . . . . . . . . . . . . . . . . . . . . . . . 73
   *Video Capturing* . . . . . . . . . . . . . . . . . . . . . . . . . . . . . . . . *73*
   *Film to Video Conversion* . . . . . . . . . . . . . . . . . . . . . . . . *75*
  VIDEO COMPRESSION . . . . . . . . . . . . . . . . . . . . . . . . . . . . 76
   *Spatial Compression (Image Compression)* . . . . . . . . . . *76*
   *Time Compression (Temporal Compression)* . . . . . . . . . *78*
   *Coding Redundancy (Data Compression)* . . . . . . . . . . . *79*
   *Pixels* . . . . . . . . . . . . . . . . . . . . . . . . . . . . . . . . . . . . . . . . . *81*
   *Blocks* . . . . . . . . . . . . . . . . . . . . . . . . . . . . . . . . . . . . . . . . *81*
   *Macroblocks* . . . . . . . . . . . . . . . . . . . . . . . . . . . . . . . . . . . *81*
   *Slice* . . . . . . . . . . . . . . . . . . . . . . . . . . . . . . . . . . . . . . . . . . *82*
   *Frames* . . . . . . . . . . . . . . . . . . . . . . . . . . . . . . . . . . . . . . . . *82*
   *Groups of Pictures (GOP)* . . . . . . . . . . . . . . . . . . . . . . . . *83*
   *Compression Scalability* . . . . . . . . . . . . . . . . . . . . . . . . . *86*
   *Advanced Video Coding (AVC/H.264)* . . . . . . . . . . . . . . *86*

IP Video Transmission . . . . . . . . . . . . . . . . . . . . . . . 88
   *File Downloading* . . . . . . . . . . . . . . . . . . . . . . *88*
   *Video Streaming* . . . . . . . . . . . . . . . . . . . . . . *89*
   *Progressive Downloading* . . . . . . . . . . . . . . . . . *91*
Digital Video Quality (DVQ) . . . . . . . . . . . . . . . . . 91
   *Tiling* . . . . . . . . . . . . . . . . . . . . . . . . . . . . . *91*
   *Error Blocks* . . . . . . . . . . . . . . . . . . . . . . . . *92*
   *Jerkiness* . . . . . . . . . . . . . . . . . . . . . . . . . . *92*
   *Artifacts* . . . . . . . . . . . . . . . . . . . . . . . . . . . *93*
   *Object Retention* . . . . . . . . . . . . . . . . . . . . . *93*
Streaming Control Protocols . . . . . . . . . . . . . . . . . 94
   *Real Time Streaming Protocol (RTSP)* . . . . . . . . . . *95*
   *Digital Storage Media Command and Control (DSM-CC)* . . . *95*
Video Formats . . . . . . . . . . . . . . . . . . . . . . . . . . 95
   *MPEG* . . . . . . . . . . . . . . . . . . . . . . . . . . . . *95*
   *Quicktime* . . . . . . . . . . . . . . . . . . . . . . . . . *96*
   *Real Media* . . . . . . . . . . . . . . . . . . . . . . . . . *96*
   *Motion JPEG (MJPEG)* . . . . . . . . . . . . . . . . . . *97*
   *Windows Media (VC-1)* . . . . . . . . . . . . . . . . . . *97*

**MOTION PICTURE EXPERTS GROUP (MPEG)** . . . . . . . . . 131
   *Digital Audio* . . . . . . . . . . . . . . . . . . . . . . . . *135*
   *Digital Video* . . . . . . . . . . . . . . . . . . . . . . . . *139*
Distribution Systems . . . . . . . . . . . . . . . . . . . . . 141
Media Streams . . . . . . . . . . . . . . . . . . . . . . . . . 142
   *Elementary Stream (ES)* . . . . . . . . . . . . . . . . . . *143*
   *Packet Elementary Stream (PES)* . . . . . . . . . . . . . *144*
   *Program Stream (PS)* . . . . . . . . . . . . . . . . . . . *144*
   *Transport Stream (TS)* . . . . . . . . . . . . . . . . . . . *145*
MPEG Transmission . . . . . . . . . . . . . . . . . . . . . . 147
   *Packet Sequencing* . . . . . . . . . . . . . . . . . . . . . *148*
   *Channel Multiplexing* . . . . . . . . . . . . . . . . . . . *150*
   *Statistical Multiplexing* . . . . . . . . . . . . . . . . . . *151*

MPEG PROGRAM TABLES . . . . . . . . . . . . . . . . . . . . . . . . . . . 152
   *Program Allocation Table (PAT)* . . . . . . . . . . . . . . . . *153*
   *Program Map Table (PMT)* . . . . . . . . . . . . . . . . . . . . *153*
   *Conditional Access Table (CAT)* . . . . . . . . . . . . . . . *153*
   *Private Tables* . . . . . . . . . . . . . . . . . . . . . . . . . . *154*
VIDEO MODES . . . . . . . . . . . . . . . . . . . . . . . . . . . . . . . . . . . 155
   *Frame Mode* . . . . . . . . . . . . . . . . . . . . . . . . . . . . *155*
   *Field Mode* . . . . . . . . . . . . . . . . . . . . . . . . . . . . . *155*
   *Mixed Mode* . . . . . . . . . . . . . . . . . . . . . . . . . . . . *155*
MEDIA FLOW CONTROL . . . . . . . . . . . . . . . . . . . . . . . . . . . 156
   *Quantizer Scaling* . . . . . . . . . . . . . . . . . . . . . . . . *156*
   *Bit Rate Control* . . . . . . . . . . . . . . . . . . . . . . . . . *157*
   *Buffering* . . . . . . . . . . . . . . . . . . . . . . . . . . . . . . *159*
   *Digital Storage Media Command and Control (DSM-CC)* . . . *160*
   *Real Time Interface (RTI)* . . . . . . . . . . . . . . . . . . . *160*
MEDIA SYNCHRONIZATION . . . . . . . . . . . . . . . . . . . . . . . . 160
DISPLAY FORMATTING . . . . . . . . . . . . . . . . . . . . . . . . . . . 161
MPEG-1 . . . . . . . . . . . . . . . . . . . . . . . . . . . . . . . . . . . . . . 162
MPEG-2 . . . . . . . . . . . . . . . . . . . . . . . . . . . . . . . . . . . . . . 163
MPEG-4 . . . . . . . . . . . . . . . . . . . . . . . . . . . . . . . . . . . . . . 164
MPEG-7 . . . . . . . . . . . . . . . . . . . . . . . . . . . . . . . . . . . . . . 166
MPEG-21 . . . . . . . . . . . . . . . . . . . . . . . . . . . . . . . . . . . . . 166
MPEG PROFILES . . . . . . . . . . . . . . . . . . . . . . . . . . . . . . . 167
   *MPEG-2 Profiles* . . . . . . . . . . . . . . . . . . . . . . . . . *168*
   *MPEG-4 Profiles* . . . . . . . . . . . . . . . . . . . . . . . . . *170*
MPEG LEVELS . . . . . . . . . . . . . . . . . . . . . . . . . . . . . . . . . 177
CONFORMANCE POINTS . . . . . . . . . . . . . . . . . . . . . . . . . . 178

IP TELEVISION SYSTEMS . . . . . . . . . . . . . . . . . . . . . . . 179

MANAGED IP TELEVISION SYSTEMS . . . . . . . . . . . . . . . . . . 179
INTERNET TELEVISION SERVICE PROVIDERS (ITVSPs) . . . . . . . . 179
PRIVATE IP TELEVISION SYSTEMS . . . . . . . . . . . . . . . . . . . 181
   *IP Television (IPTV) Networks* . . . . . . . . . . . . . . . . *181*

CONTRIBUTION NETWORK . . . . . . . . . . . . . . . . . . . . . . 183
    *Connection Types* . . . . . . . . . . . . . . . . . . . . . . . . . . .*183*
    *Program Transfer Scheduling* . . . . . . . . . . . . . . . . . . . . .*186*
    *Content Feeds* . . . . . . . . . . . . . . . . . . . . . . . . . . . .*187*
HEADEND . . . . . . . . . . . . . . . . . . . . . . . . . . . . . . 190
    *Integrated Receiver Decoder (IRD)* . . . . . . . . . . . . . . . . . .*191*
    *Off Air Receivers* . . . . . . . . . . . . . . . . . . . . . . . . . .*192*
    *Encoders* . . . . . . . . . . . . . . . . . . . . . . . . . . . . . .*193*
    *Transcoders* . . . . . . . . . . . . . . . . . . . . . . . . . . . .*193*
    *Packet Switch* . . . . . . . . . . . . . . . . . . . . . . . . . . .*193*
ASSET MANAGEMENT . . . . . . . . . . . . . . . . . . . . . . . 194
    *Content Acquisition* . . . . . . . . . . . . . . . . . . . . . . . . .*195*
    *Metadata Management* . . . . . . . . . . . . . . . . . . . . . . .*195*
    *Playout Scheduling* . . . . . . . . . . . . . . . . . . . . . . . . .*196*
    *Asset Storage* . . . . . . . . . . . . . . . . . . . . . . . . . . .*197*
    *Content Processing* . . . . . . . . . . . . . . . . . . . . . . . . .*198*
    *Ad Insertion* . . . . . . . . . . . . . . . . . . . . . . . . . . . .*199*
    *Distribution Control* . . . . . . . . . . . . . . . . . . . . . . . .*201*
DISTRIBUTION NETWORK . . . . . . . . . . . . . . . . . . . . . 202
    *Core Network* . . . . . . . . . . . . . . . . . . . . . . . . . . . .*203*
ACCESS NETWORK . . . . . . . . . . . . . . . . . . . . . . . . . 204
    *Digital Subscriber Line (DSL)* . . . . . . . . . . . . . . . . . . . .*205*
    *Cable Modem* . . . . . . . . . . . . . . . . . . . . . . . . . . . .*207*
    *Wireless Broadband* . . . . . . . . . . . . . . . . . . . . . . . .*209*
    *Power Line Carrier (PLC)* . . . . . . . . . . . . . . . . . . . . . .*214*
    *Optical Access Network* . . . . . . . . . . . . . . . . . . . . . . .*215*
PREMISES DISTRIBUTION . . . . . . . . . . . . . . . . . . . . . . 216


**PREMISES DISTRIBUTION NETWORKS (PDN)** . . . . . . . . **221**

HOME MULTIMEDIA SERVICE NEEDS . . . . . . . . . . . . . . . . 222
    *Telephone* . . . . . . . . . . . . . . . . . . . . . . . . . . . . . .*223*
    *Internet Access* . . . . . . . . . . . . . . . . . . . . . . . . . . .*223*

*Television* . . . . . . . . . . . . . . . . . . . . . . . . . . . . . . . . . *224*
*Interactive Video* . . . . . . . . . . . . . . . . . . . . . . . . . . . . *224*
*Media Streaming* . . . . . . . . . . . . . . . . . . . . . . . . . . . . *224*
HOME MULTIMEDIA SYSTEM NEEDS . . . . . . . . . . . . . . . . . 226
*Co-Existence* . . . . . . . . . . . . . . . . . . . . . . . . . . . . . . . *226*
*No New Wires* . . . . . . . . . . . . . . . . . . . . . . . . . . . . . . *227*
*Data Rates* . . . . . . . . . . . . . . . . . . . . . . . . . . . . . . . . *228*
*Quality of Service (QoS)* . . . . . . . . . . . . . . . . . . . . . . *228*
*Home Coverage* . . . . . . . . . . . . . . . . . . . . . . . . . . . . . *229*
*Security* . . . . . . . . . . . . . . . . . . . . . . . . . . . . . . . . . . *229*
*Cost* . . . . . . . . . . . . . . . . . . . . . . . . . . . . . . . . . . . . . *230*
*Installation* . . . . . . . . . . . . . . . . . . . . . . . . . . . . . . . *230*
HOME NETWORKING TECHNOLOGIES . . . . . . . . . . . . . . . . 231
*Adaptive Modulation* . . . . . . . . . . . . . . . . . . . . . . . . *232*
*Echo Control* . . . . . . . . . . . . . . . . . . . . . . . . . . . . . . *232*
*Synchronized Transmission* . . . . . . . . . . . . . . . . . . . *233*
*Interference Avoidance* . . . . . . . . . . . . . . . . . . . . . . *234*
*Power Level Control* . . . . . . . . . . . . . . . . . . . . . . . . *234*
*Channel Bonding* . . . . . . . . . . . . . . . . . . . . . . . . . . . *235*
TRANSMISSION TYPES . . . . . . . . . . . . . . . . . . . . . . . . . . . 236
*Wired LAN* . . . . . . . . . . . . . . . . . . . . . . . . . . . . . . . . *236*
*Wireless* . . . . . . . . . . . . . . . . . . . . . . . . . . . . . . . . . . *237*
*Power Line* . . . . . . . . . . . . . . . . . . . . . . . . . . . . . . . . *239*
*Coaxial* . . . . . . . . . . . . . . . . . . . . . . . . . . . . . . . . . . *242*
*Phoneline* . . . . . . . . . . . . . . . . . . . . . . . . . . . . . . . . *245*
PREMISES DISTRIBUTION SYSTEMS . . . . . . . . . . . . . . . . . 247
*HomePlug™* . . . . . . . . . . . . . . . . . . . . . . . . . . . . . . . *247*
*HomePNA™* . . . . . . . . . . . . . . . . . . . . . . . . . . . . . . . *254*
*Digital Home Standard (DHS)* . . . . . . . . . . . . . . . . . *260*
*HD-PLC* . . . . . . . . . . . . . . . . . . . . . . . . . . . . . . . . . *261*
*TVnet* . . . . . . . . . . . . . . . . . . . . . . . . . . . . . . . . . . . *261*
*Multimedia over Coax Alliance (MoCA)™* . . . . . . . . . . *264*
*802.11 Wireless LAN* . . . . . . . . . . . . . . . . . . . . . . . . *266*

**IPTV END USER DEVICES** . . . . . . . . . . . . . . . . . . . . . . . . . . **273**

 *Set Top Boxes (STB)* . . . . . . . . . . . . . . . . . . . . . . . . . *274*
 *Internet Protocol Set Top Box (IP STB)* . . . . . . . . . . . . . *275*
 *Download and Play (DP STB)* . . . . . . . . . . . . . . . . . . . *275*
 *Hybrid STB* . . . . . . . . . . . . . . . . . . . . . . . . . . . . . . . . *275*
 *IP Television (IP Television)* . . . . . . . . . . . . . . . . . . . . *276*
 *Mobile Video Telephones* . . . . . . . . . . . . . . . . . . . . . . . *276*
 *Multimedia Computers* . . . . . . . . . . . . . . . . . . . . . . . . . *276*
 *Portable Media Players* . . . . . . . . . . . . . . . . . . . . . . . . *277*
 IP STB CAPABILITIES . . . . . . . . . . . . . . . . . . . . . . . . . . . 277
 *Network Connection Types* . . . . . . . . . . . . . . . . . . . . . . *278*
 *Display Capability* . . . . . . . . . . . . . . . . . . . . . . . . . . . *278*
 *Video Scaling* . . . . . . . . . . . . . . . . . . . . . . . . . . . . . . . *278*
 *Dual Display (Picture in Picture)* . . . . . . . . . . . . . . . . *278*
 *Security* . . . . . . . . . . . . . . . . . . . . . . . . . . . . . . . . . . . *279*
 *Smart Card* . . . . . . . . . . . . . . . . . . . . . . . . . . . . . . . . *279*
 *DRM Client* . . . . . . . . . . . . . . . . . . . . . . . . . . . . . . . . *279*
 *Secure Microprocessor* . . . . . . . . . . . . . . . . . . . . . . . . *280*
 *Media Processing* . . . . . . . . . . . . . . . . . . . . . . . . . . . . *280*
 *Communication Protocols* . . . . . . . . . . . . . . . . . . . . . . *281*
 *Software Applications* . . . . . . . . . . . . . . . . . . . . . . . . . *282*
 *Accessories* . . . . . . . . . . . . . . . . . . . . . . . . . . . . . . . . . *282*
 *Middleware Compatibility* . . . . . . . . . . . . . . . . . . . . . . *282*
 *Upgradability* . . . . . . . . . . . . . . . . . . . . . . . . . . . . . . . *283*
 *Plug-In* . . . . . . . . . . . . . . . . . . . . . . . . . . . . . . . . . . . . *283*
 *Media Portability* . . . . . . . . . . . . . . . . . . . . . . . . . . . . *283*
 END USER DEVICE OPERATION . . . . . . . . . . . . . . . . . . . . 283
 *Network Interface* . . . . . . . . . . . . . . . . . . . . . . . . . . . . *284*
 *Network Selection* . . . . . . . . . . . . . . . . . . . . . . . . . . . . *285*
 *Channel Selection* . . . . . . . . . . . . . . . . . . . . . . . . . . . . *285*
 *Return Path* . . . . . . . . . . . . . . . . . . . . . . . . . . . . . . . . *285*
 *Demultiplexing* . . . . . . . . . . . . . . . . . . . . . . . . . . . . . . *285*
 *Decoding* . . . . . . . . . . . . . . . . . . . . . . . . . . . . . . . . . . . *286*
 *Rendering* . . . . . . . . . . . . . . . . . . . . . . . . . . . . . . . . . . *286*

*Video Coding* . . . . . . . . . . . . . . . . . . . . . . . . . . . . . *286*
*Audio Coding* . . . . . . . . . . . . . . . . . . . . . . . . . . . . . *286*
*Graphic Processing* . . . . . . . . . . . . . . . . . . . . . . . . . *287*
*User Interface* . . . . . . . . . . . . . . . . . . . . . . . . . . . . *287*
*On Screen Display (OSD)* . . . . . . . . . . . . . . . . . . . . *287*
*Electronic Programming Guide (EPG)* . . . . . . . . . . . . . *288*
*Remote Control* . . . . . . . . . . . . . . . . . . . . . . . . . . . *288*
*Wireless Keyboard* . . . . . . . . . . . . . . . . . . . . . . . . . *288*
*Infrared Blaster* . . . . . . . . . . . . . . . . . . . . . . . . . . *288*
*Web Interface* . . . . . . . . . . . . . . . . . . . . . . . . . . . . *289*
*Game Port* . . . . . . . . . . . . . . . . . . . . . . . . . . . . . . *289*
*Universal Serial Bus (USB)* . . . . . . . . . . . . . . . . . . . *289*
PREMISES DISTRIBUTION . . . . . . . . . . . . . . . . . . . . . 289
INTERFACES . . . . . . . . . . . . . . . . . . . . . . . . . . . . . . 290
*Network Connections* . . . . . . . . . . . . . . . . . . . . . . . *290*
*Ethernet* . . . . . . . . . . . . . . . . . . . . . . . . . . . . . . . *290*
*Digital Subscriber Line (DSL)* . . . . . . . . . . . . . . . . . *290*
*Satellite Connection (Optional)* . . . . . . . . . . . . . . . . *291*
*DTT Connection (Optional)* . . . . . . . . . . . . . . . . . . . *291*
*Cable TV Connection (Optional)* . . . . . . . . . . . . . . . . *291*
*TV RF Output* . . . . . . . . . . . . . . . . . . . . . . . . . . . *291*
*Video* . . . . . . . . . . . . . . . . . . . . . . . . . . . . . . . . . *291*
*Composite Video* . . . . . . . . . . . . . . . . . . . . . . . . . . *292*
*Component Video* . . . . . . . . . . . . . . . . . . . . . . . . . *292*
*S-Video* . . . . . . . . . . . . . . . . . . . . . . . . . . . . . . . *292*
*SCART* . . . . . . . . . . . . . . . . . . . . . . . . . . . . . . . . *293*
*IEEE 1394* . . . . . . . . . . . . . . . . . . . . . . . . . . . . . *294*
*High Definition Multimedia Interface (HDMI)* . . . . . . . . *295*
*Audio* . . . . . . . . . . . . . . . . . . . . . . . . . . . . . . . . . *296*
*Baseband Audio* . . . . . . . . . . . . . . . . . . . . . . . . . . *296*
*Sony Philips Digital Interface (S/PDIF)* . . . . . . . . . . . *296*
*Wireless* . . . . . . . . . . . . . . . . . . . . . . . . . . . . . . . *296*
*Wireless LAN (WLAN)* . . . . . . . . . . . . . . . . . . . . . . *296*
*Ultra Wideband (UWB)* . . . . . . . . . . . . . . . . . . . . . *297*

# DIGITAL RIGHTS MANAGEMENT (DRM) . . . . . . . . . . . . . . 299

INTELLECTUAL PROPERTY RIGHTS . . . . . . . . . . . . . . . . . . . . . . 302
   *Copyright* . . . . . . . . . . . . . . . . . . . . . . . . . . . . . . . . . . . . . . *303*
   *Patents* . . . . . . . . . . . . . . . . . . . . . . . . . . . . . . . . . . . . . . . *303*
   *Trademarks* . . . . . . . . . . . . . . . . . . . . . . . . . . . . . . . . . . . . *304*
   *Trade Secrets (Confidential Information)* . . . . . . . . . . . . . . . *304*
TYPES OF RIGHTS . . . . . . . . . . . . . . . . . . . . . . . . . . . . . . . . . . 305
   *Rendering Rights* . . . . . . . . . . . . . . . . . . . . . . . . . . . . . . . . *305*
   *Transport Rights* . . . . . . . . . . . . . . . . . . . . . . . . . . . . . . . . *306*
   *Derivative Rights* . . . . . . . . . . . . . . . . . . . . . . . . . . . . . . . . *306*
DIGITAL RIGHTS MANAGEMENT . . . . . . . . . . . . . . . . . . . . . . . 308
   *Rights Attributes* . . . . . . . . . . . . . . . . . . . . . . . . . . . . . . . . *309*
   *Rights Models* . . . . . . . . . . . . . . . . . . . . . . . . . . . . . . . . . . *309*
DIGITAL ASSETS . . . . . . . . . . . . . . . . . . . . . . . . . . . . . . . . . . . 314
   *Digital Audio (Voice and Music)* . . . . . . . . . . . . . . . . . . . . *315*
   *Images* . . . . . . . . . . . . . . . . . . . . . . . . . . . . . . . . . . . . . . . . *317*
   *Animation* . . . . . . . . . . . . . . . . . . . . . . . . . . . . . . . . . . . . . *317*
   *Digital Video* . . . . . . . . . . . . . . . . . . . . . . . . . . . . . . . . . . *318*
   *Data Files (Books and Databases)* . . . . . . . . . . . . . . . . . . . *319*
   *Application Program Files (Software Applications)* . . . . . . . *319*
   *Container Files* . . . . . . . . . . . . . . . . . . . . . . . . . . . . . . . . . *319*
   *Digital Asset Management (DAM)* . . . . . . . . . . . . . . . . . . . *321*
MEDIA IDENTIFICATION . . . . . . . . . . . . . . . . . . . . . . . . . . . . . 324
   *Unique Material Identifier (UMID)* . . . . . . . . . . . . . . . . . . *324*
   *International Standard Book Number (ISBN)* . . . . . . . . . . . *325*
   *International Standard Serial Number (ISSN)* . . . . . . . . . . . *326*
   *Digital Object Identifier (DOI)* . . . . . . . . . . . . . . . . . . . . . *326*
   *International Standard Recording Code (ISRC)* . . . . . . . . . . *327*
   *International Standard Audiovisual Number (ISAN)* . . . . . . *327*
   *International Standard Work Code (ISWC)* . . . . . . . . . . . . . *328*
DRM SECURITY PROCESSES . . . . . . . . . . . . . . . . . . . . . . . . . . 328
   *Authentication* . . . . . . . . . . . . . . . . . . . . . . . . . . . . . . . . . . *328*
   *Encryption* . . . . . . . . . . . . . . . . . . . . . . . . . . . . . . . . . . . . . *331*
   *Digital Watermarks* . . . . . . . . . . . . . . . . . . . . . . . . . . . . . . *336*

*Digital Fingerprint* . . . . . . . . . . . . . . . . . . . . . . . . . . . .*337*
*Digital Certificate* . . . . . . . . . . . . . . . . . . . . . . . . . . . . .*338*
*Digital Signature* . . . . . . . . . . . . . . . . . . . . . . . . . . . . . .*339*
*Secure Hypertext Transfer Protocol (S-HTTP)* . . . . . . . . . .*340*
*Machine Binding* . . . . . . . . . . . . . . . . . . . . . . . . . . . . . .*340*
*Conditional Access (CA)* . . . . . . . . . . . . . . . . . . . . . . . . .*340*
*Product Activation* . . . . . . . . . . . . . . . . . . . . . . . . . . . . .*340*
*Secure Socket Layer (SSL)* . . . . . . . . . . . . . . . . . . . . . . .*341*
*Transport Layer Security (TLS)* . . . . . . . . . . . . . . . . . . .*342*
DRM SYSTEM . . . . . . . . . . . . . . . . . . . . . . . . . . . . . . . . 342
*Content Server* . . . . . . . . . . . . . . . . . . . . . . . . . . . . . . .*344*
*Metadata* . . . . . . . . . . . . . . . . . . . . . . . . . . . . . . . . . . .*344*
*DRM Packager* . . . . . . . . . . . . . . . . . . . . . . . . . . . . . . .*345*
*License Server* . . . . . . . . . . . . . . . . . . . . . . . . . . . . . . .*345*
*Key Server* . . . . . . . . . . . . . . . . . . . . . . . . . . . . . . . . . .*345*
*DRM Controller* . . . . . . . . . . . . . . . . . . . . . . . . . . . . . .*346*
*DRM Client* . . . . . . . . . . . . . . . . . . . . . . . . . . . . . . . . .*346*
MEDIA TRANSFER . . . . . . . . . . . . . . . . . . . . . . . . . . . . . 346
*File Downloading* . . . . . . . . . . . . . . . . . . . . . . . . . . . . . .*347*
*Media Streaming* . . . . . . . . . . . . . . . . . . . . . . . . . . . . . .*348*
*Stored Media Distribution* . . . . . . . . . . . . . . . . . . . . . . .*348*
MEDIA DISTRIBUTION . . . . . . . . . . . . . . . . . . . . . . . . . . 348
*Direct Distribution* . . . . . . . . . . . . . . . . . . . . . . . . . . . .*349*
*Superdistribution* . . . . . . . . . . . . . . . . . . . . . . . . . . . . .*349*
*Peer to Peer Distribution* . . . . . . . . . . . . . . . . . . . . . . . .*349*
*Media Portability* . . . . . . . . . . . . . . . . . . . . . . . . . . . . .*349*
KEY MANAGEMENT . . . . . . . . . . . . . . . . . . . . . . . . . . . . 349
*Smart Card* . . . . . . . . . . . . . . . . . . . . . . . . . . . . . . . . .*350*
*Virtual Card* . . . . . . . . . . . . . . . . . . . . . . . . . . . . . . . .*351*
DIGITAL RIGHTS MANAGEMENT THREATS . . . . . . . . . . . . . 351
*Ripping* . . . . . . . . . . . . . . . . . . . . . . . . . . . . . . . . . . . .*351*
*Hacking* . . . . . . . . . . . . . . . . . . . . . . . . . . . . . . . . . . . .*352*
*Spoofing* . . . . . . . . . . . . . . . . . . . . . . . . . . . . . . . . . . . .*353*
*Hijacking* . . . . . . . . . . . . . . . . . . . . . . . . . . . . . . . . . . .*353*
*Bit Torrent* . . . . . . . . . . . . . . . . . . . . . . . . . . . . . . . . . .*354*

*Camming* . . . . . . . . . . . . . . . . . . . . . . . . . . . . . . . . *355*
*Insider Piracy* . . . . . . . . . . . . . . . . . . . . . . . . . . . *355*
*Analog Hole* . . . . . . . . . . . . . . . . . . . . . . . . . . . . . *355*
*Digital Hole* . . . . . . . . . . . . . . . . . . . . . . . . . . . . . *356*
*Misrouting* . . . . . . . . . . . . . . . . . . . . . . . . . . . . . . *356*
PROTOCOLS AND INDUSTRY STANDARDS . . . . . . . . . . . . . . . . . . . 356
*Extensible Rights Management Language (XrML)* . . . . . . . . *357*
*Extensible Media Commerce Language (XMCL)* . . . . . . . . . . *358*
*Open Digital Rights Language (ODRL)* . . . . . . . . . . . . . *358*
*MPEG-7* . . . . . . . . . . . . . . . . . . . . . . . . . . . . . . . . *358*
*MPEG-21* . . . . . . . . . . . . . . . . . . . . . . . . . . . . . . . *358*
*Contracts Expression Language (CEL)* . . . . . . . . . . . . . *359*
*Secure Digital Music Initiative (SDMI)* . . . . . . . . . . . . *359*
*Resource Description Framework (RDF)* . . . . . . . . . . . . . *359*
*World Wide Web consortium (W3C).* . . . . . . . . . . . . . . . . *360*
COPY PROTECTION SYSTEMS . . . . . . . . . . . . . . . . . . . . . . . . 360
*Copy Control Information (CCI)* . . . . . . . . . . . . . . . . . *361*
*Copy Generation Management System (CGMS)* . . . . . . . . . . . *361*
*Broadcast Flag* . . . . . . . . . . . . . . . . . . . . . . . . . . . *361*
*Serial Copy Management System (SCMS)* . . . . . . . . . . . . . *362*
*Content Protection for Prerecorded Media (CPPM)* . . . . . . . *362*
*Content Protection for Recordable Media (CPRM)* . . . . . . . . *362*
*Content Scrambling System (CSS)* . . . . . . . . . . . . . . . . *362*
*Secure Video Processing (SVP)* . . . . . . . . . . . . . . . . . *363*
*Digital Transmission Content Protection (DTCP)* . . . . . . . . *363*
*High Bandwidth Digital Content Protection (HDCP)* . . . . . . . *363*
*High Definition Multimedia Interface (HDMI)* . . . . . . . . . . *363*
*Extended Conditional Access (XCA)* . . . . . . . . . . . . . . . *364*

**APPENDIX 1 - ACRONYMS** ........................... 365

**APPENDIX 2 - STANDARDS** ......................... 375

**INDEX** ............................................ 379

# Chapter 1

## Internet Protocol Television(IPTV)

IPTV is a process of sending television signals over IP data networks . These IP networks can be managed (e.g. DSL or Optical) or they can be unmanaged (e.g. broadband Internet). If the television signal is in analog form (standard TV or HDTV) the video and audio signals are first converted to a digital form. Packet routing information is then added to the digital video and voice signals so they can be routed through the Internet or data network.

The viewing devices or adapters convert digital television signals into a form that can be controlled and viewed by users. Broadband access providers supply the high-speed data connection that can transfer the digital video television signals. Service providers identify and control the connections between the viewing devices and the content providers (media sources). Media content providers create information that people want to view or obtain.

To get IPTV you need these key parts:

- Viewing devices or adapters
- Broadband access providers
- IPTV service providers
- Media content providers

The key types of IPTV viewing devices include multimedia computers, television adapter boxes, multimedia mobile telephones and IPTVs. Multimedia computers have video processing and audio processing capabilities. Television adapters convert digital television signals into standard television RF connections that allow standard televisions to watch IPTV channels. IP televisions are devices that are specifically designed to watch television channels through IP data networks without the need for adapter boxes or media gateways.

Figure 1.1 shows several types of IPTV viewing devices. This diagram shows that some of the options for viewing devices include multimedia computers, television adapters, IP televisions and mobile telephones. Multimedia computers (desktops and laptops) allow some viewers to watch Internet television programs without the need for adapters provided they have the multimedia browsers that have the appropriate media plug-ins. Television adapters (set top boxes) connect standard television RF or video and audio

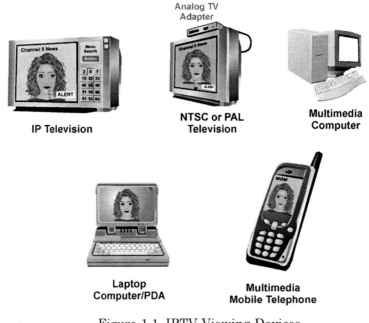

Figure 1.1, IPTV Viewing Devices

connectors to data jacks or wireless LAN connections. IP televisions can be directly connected to data jacks or wireless LAN connections. Mobile telephones that have multimedia play capabilities along with broadband wireless services and the necessary media gateways.

Broadband access providers transfer high-speed data to the end users. The type of technology used by broadband access providers can play an important part in the ability and/or quality of IPTV services.

Figure 1.2 shows the key types of broadband access providers that can be used to provide Internet television service. This diagram shows that some of the common types of broadband access systems that are available include powerline data distribution, cable modems, digital subscriber lines (DSL) and wireless local area network systems (3G wireless, WLAN, MMDS, and LMDS).

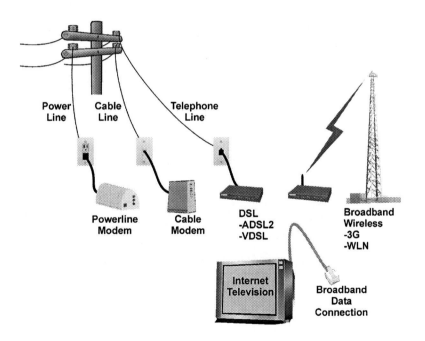

Figure 1.2, Broadband Access Providers

Internet television service providers (ITVSPs) help customers find Internet television channels and manage connections between media sources. While it is possible in some cases for end users to directly connect to a media source by using an IP address or even a web link, Internet television provides may simplify (and limit) the programming guide choices. Internet television service providers may also provide connections to subscription controlled television sources. For this role, the IPTV service provider makes a business relationship with the media source. The IPTV service provider may pay the content provider a fixed fee or share the revenue from funds it collects from their end user.

Figure 1.3 shows that Internet television service providers (ITVSPs) are primarily made of computers that are connected to the Internet and software to operate call processing and other services. In this diagram, a computer keeps track of which customers are active (registration) and what features and services are authorized. When television channel requests are processed, the ITVSP sends messages to gateways via the Internet allowing television channels to be connected to a selected media gateway source (such as television channels). These media gateways transfer their billing details to a clearinghouse so the ITVSP can pay for the gateway's usage. The ITVSP then can use this billing information to charge the customer for channels viewed.

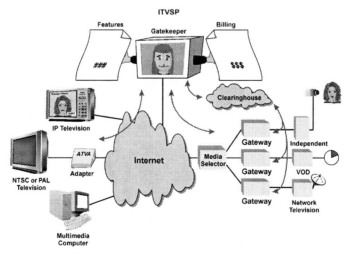

Figure 1.3, IPTV Service Providers

Content providers create different types of television media programs that are distributed through a variety of media distribution channels. Common program distribution channels include movie theaters, pay per view, airlines and specialty channels, video rentals, and television networks. Traditional television network content sources include movies, television programs, sports channels, news services and other information sources.

Television networks have traditionally been a closed distribution system where the television network determined which program sources could reach viewers. Television networks such as cable television, satellite systems and VHF/UHF transmission systems have a limited number of channels (up to several hundred channels).

IPTV systems can be provided through broadband communication systems that can reach content providers and viewers in any part of the world. This allows IPTV service providers to offer many new content programs that have not been available for standard television distribution systems. New types of content sources include personal media channels, global television channels, interactive media, public video sources and private video sources.

Personal media channels allow viewers to create their own television channel, upload their content (such as pictures and videos) and share their content with other IPTV viewers. IPTV systems provide access to television channels throughout the world. Some of the more popular global television channels that are available on IPTV include news channels, business channels and music television. IPTVs may provide access to interactive media such as games, chat rooms and e-commerce shopping. Governments and public groups have begun providing real time video access to public sources such as courtrooms, popular public places and public web cams. Private video sources are provided by companies or people who are willing to provide and pay to have their signals available on television. Some common private television channels include religious groups, education sources and sports channels.

Figure 1.4 shows some of the existing and new types of IPTV content providers. This diagram shows that IPTV content includes traditional television content sources such as movies, news services, sports, education, religious and other forms of one-way information content. New types of IPTV content include personal media, global television channels, interactive media, public video sources and private video sources.

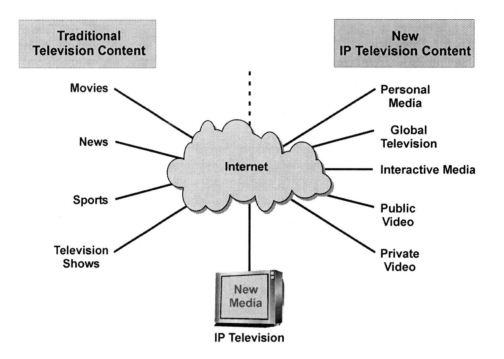

Figure 1.4, IPTV Media Content Providers

## Why Consider IPTV and Internet Television Services

There are three key reasons why people and companies are adding to or converting their existing television systems to television over data network capabilities:

1. More channels
2. More control
3. More services

## More Channels

To provide IPTV service, a service provider only needs to manage one connection between the end viewer and the media source (such as a television channel). IPTV service providers simply connect (through a switch) to the media source that the viewer has selected at their distribution location. Because IPTV channels can be routed through data networks (such as the Internet), this allows an IPTV service provider to offer thousands or tens of thousands of channels.

IPTV systems can connect viewers to media sources and the media sources can be located anywhere in the world. All that is needed to provide high-quality television channels is a media server, a 2 Mbps to 4 Mbps data connection and a viewing device. Some channels may be provided free of charge while others require payment for viewing options.

Figure 1.5 shows that Internet television offers virtually an unlimited number of media channels. This example shows an Internet television service provider that offers access to more than 10,000 media channels. The viewer is first provided with the option to select which country to select. Once a country has been selected, the available Internet television channels are shown along with the cost of accessing these channels. For some selections, the channel is free and for other selections, the user will be charged a fee (units of service in this example). This user has 148 units remaining for viewing channels this month.

## More Control

IPTV viewers can have more control over their television services. They can choose from multiple service providers rather than a single television service provider that is located in their area. IPTV systems may allow the customer to setup their own IPTV service (instant activation) through a web page. The customer can add and remove features when they desire.

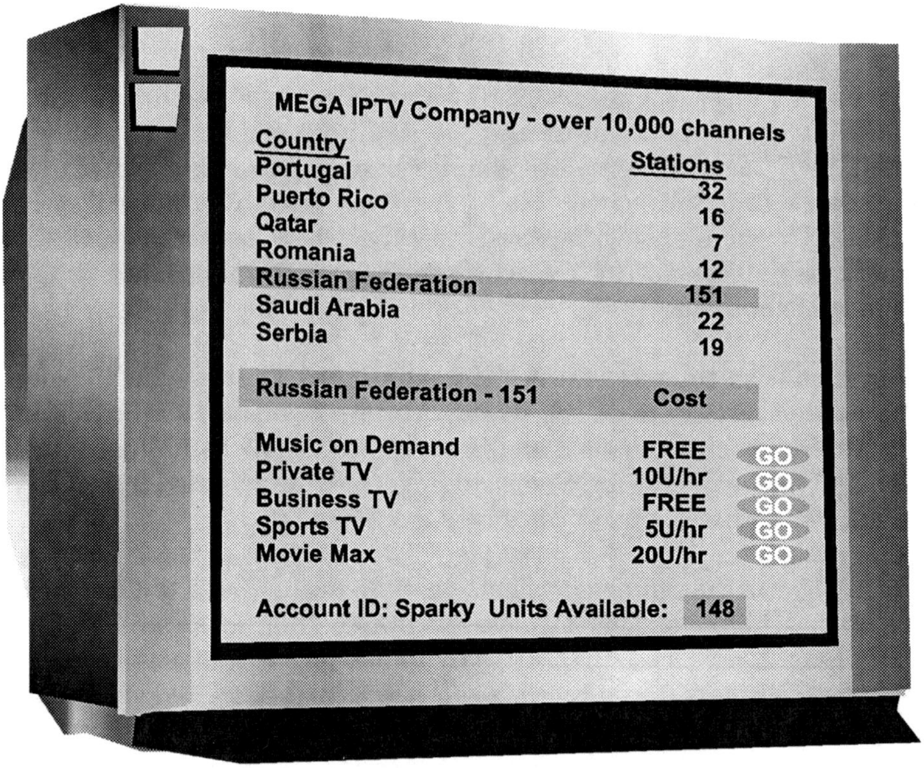

Figure 1.5, More IPTV Channels

Because IPTV has the capability of having multiple channels, it is possible to have multiple channels for each media program allowing intelligent filtering of content such as restricting channels with unacceptable video or audio.

Figure 1.6 shows how IPTV allows viewers to have more control over their television services. This diagram shows that the customer can select from several alternate service providers and Internet television systems have the capability of allowing new viewers to instantly subscribe to services and select features. This example shows that a local telephone company offers television over a managed DSL connection which provides a guaranteed quality of service and access to local television channels. The Cheap TV provider offers access to many free channels throughout the world. The Premier TV company provides high quality television and access to movie channels.

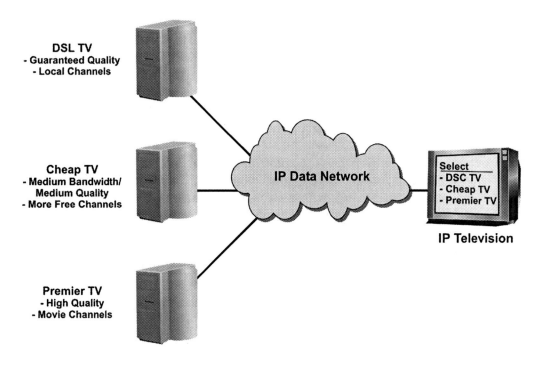

Figure 1.6, More Control over Television Services

## More Services

Television service is the providing of television information to customers by a common carrier, administration, or private operating agency using voice, data, and/or video technologies. IPTV systems have new capabilities that allow service providers to offer new services to viewers.

IPTV systems deliver information that has been converted to a standard digital form. This allows IPTV systems to individually address video sources to specific users and to allow for interactive two-way services. Some of the advanced services made possible for IPTV include addressable advertising messages, interactive multiplayer gaming, advanced electronic programming guides, personal media channels, shopping channels, and global television channels.

Figure 1.6 shows how IPTV systems can provide new services that are not possible with traditional television systems. This example shows that almost any type of digital media can be addressed to specific television devices and the viewer can interactively respond to prompts and initiate control requests (two-way capability). This allows service providers to offer new media sources such as ads that are sent to specific viewers that have a matching profile or dynamic programs that allow the user to select or change the content that is presented to them. Viewers may be given access to public video (such as traffic video monitors) or private video (such as security video cameras at home and day care locations). Two way control services could allow users to interactively search for programs they want to view, access games and community media centers, web commerce and advertising messages that offer the viewer a selection of options such as more information or to be redirected to an infomercial. This example shows a program that allows the user to select the ending of a scene where a gunman is threatening a woman with a gun. The user can select a happy ending (the girl gets away) or a sad ending (the girl gets shot).

## How IPTV and Internet Television Systems Work

Understanding the basics of how IPTV and Internet television service works will help you make better choices and may help you to solve problems that can be caused by selecting the wrong types of technologies, equipment and services.

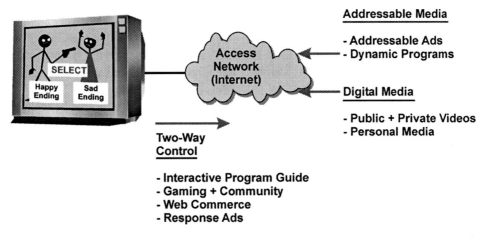

Figure 1.6, More Television Services

## Digitization - Converting Video Signals and Audio Signals to Digital Signals

A key first step in providing IPTV service is having the media in digital form. If the video is in analog form, the video and audio signals are converted into a digital form (digitization) and then compressing the digitized information into a more efficient form.

Digitization is the conversion of analog signals (continually varying signals) into digital form (signals that have only two levels). To convert analog signals to digital form, the analog signal is sampled and digitized by using an analog-to-digital (pronounced A to D) converter. The A/D converter periodically senses (samples) the level of the analog signal and creates a binary number or series of digital pulses that represent the level of the signal.

## Digital Media Compression – Gaining Efficiency

Digital media compression is a process of analyzing a digital signal (digitized video and/or audio) and using the analysis information to convert the high-speed digital signals that represent the actual signal shape into lower-speed digital signals that represent the actual content (such as a moving image or human voice). This process allows IPTV service to have lower data transmission rates than standard digital video signals while providing for good quality video and audio. Digital media compression for IPTV includes digital audio compression and digital video compression.

Figure 1.7 shows the basic digital speech compression process. In this example, the word "HELLO" is digitized. The initial digitized bits represent every specific shape of the digitized word HELLO. This digital information is analyzed and it is determined that this entire word can be represented by three sounds: "HeH" + "LeL" + "OH." Each of these sounds only requires a few digital bits instead of the many bits required to recreate the entire analog waveform.

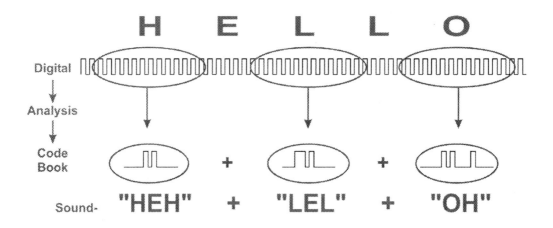

Figure 1.7, Digital Audio Compression

Figure 1.8 demonstrates the operation of the basic digital video compression system. Each video frame is digitized and then sent for digital compression. The digital compression process creates a sequence of frames (images) that start with a key frame. The key frame is digitized and used as reference points for the compression process. Between the key frames, only the differences in images are transmitted. This dramatically reduces the data transmission rate to represent a digital video signal as an uncompressed digital video signal requires over 270 Mbps compared to less than 4 Mbps for a typical digital video disk (DVD) digital video signal.

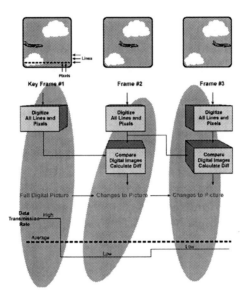

Figure 1.8, Digital Video Compression

## Sending Packets

Sending packets through the Internet involves routing them through the network and managing the loss of packets when they can't reach their destination.

Packet Routing Methods

Packet routing involves the transmission of packets through intelligent switches (called routers) that analyze the destination address of the packet and determine a path that will help the packet travel toward its destination.

Routers learn from each other about the best routes for them to select when forwarding packets toward their destination (usually paths to other routers). Routers regularly broadcast their connection information to nearby routers and they listen for connection information from neighboring routers. From this information, routers build information tables (called routing tables) that help them to determine the best path for them to forward each packet to.

Routers may forward packets toward their destination simply based on their destination address or they may look at some descriptive information about the packet. This descriptive information may include special handling instructions (called a label or tag) or priority status (such as high priority for real time voice or video signals).

Figure 1.9 shows how blocks of data are divided into small packet sizes that can be sent through the Internet. After the data is divided into packets (envelopes shown in this example), a destination address along with some description about the contents is added to each packet (called in the packet header). As the packet enters into the Internet (routing boxes shown in this

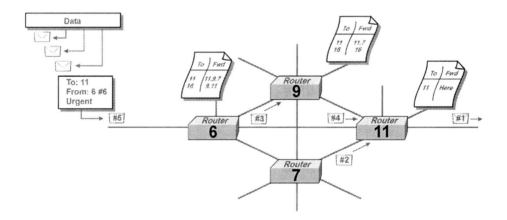

Figure 1.9, Packet Transmission

diagram), each router reviews the destination address in its routing table and determines which paths it can send the packet to so it will move further towards its destination. If a current path is busy or unavailable (such as shown for packet #3), the router can forward the packets to other routers that can forward the packet towards its destination. This example shows that because some packets will travel through different paths, packets may arrive out of sequence at their destination. When the packets arrive at their destination, they can be reassembled into proper order using the packet sequence number.

Packet Losses and Effects on Television Quality

Packet losses are the incomplete reception or intentional discarding of data packets as they are sent through a network. Packets may be lost due to broken line connections, distortion from electrical noise (e.g. from a lightning spike), or through intentional discarding due to congested switch conditions. Packet losses are usually measured by counting the number of data packets that have been lost in transmission compared to the total number of packets that have been transmitted.

Figure 1.10 shows how some packets may be lost during transmission through a communications system. This example shows that several packets enter into the Internet. The packets are forwarded toward their destination as usual. Unfortunately, a lighting strike corrupts (distorts) packet 8 and it cannot be forwarded. Packet 6 is lost (discarded) when a router has exceeded its capacity to forward packets because too many were arriving at the same time. This diagram shows that the packets are serialized to allow them to be placed in correct order at the receiving end. When the receiving end determines a packet is missing in the sequence, it can request that another packet be retransmitted. If the time delivery of packets is critical (such as for packetized video), it is common that packet retransmission requests are not performed and the lost packets simply result in distortion of the received information (such as poor video quality).

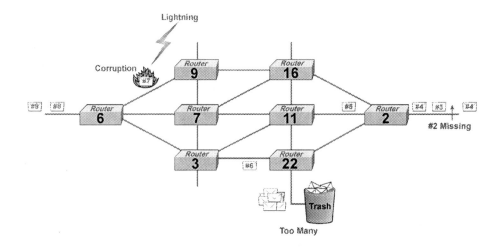

Figure 1.10, Packet Losses

Packet Buffering

Packet buffering is the process of temporarily storing (buffering) packets during the transmission of information to create a reserve of packets that can be used during packet transmission delays or retransmission requests. While a packet buffer is commonly located in the receiving device, a packet buffer may also be used in the sending device to allow the rapid selection and retransmission of packets when they are requested by the receiving device. Packet buffering is commonly used in IPTV systems to overcome the transmission delays and packet losses that occur when viewing IPTV signals.

A packet buffer receives and adds small amounts of delay to packets so that all the packets appear to have been received without varying delays. The amount of packet buffering for IPTV systems can vary from tenths of a second to tens of seconds.

Figure 1.11 shows how packet buffering can be used to reduce the effects of packet delays and packet loss for streaming media systems. This diagram shows that during the transmission of packets from the media server to the viewer, some of the packet transmission time varies (jitter) and some of the packets are lost during transmission. The packet buffer temporarily stores data before providing it to the media player. This provides the time necessary to time synchronize the packets and to request and replace packets that have been lost during transmission.

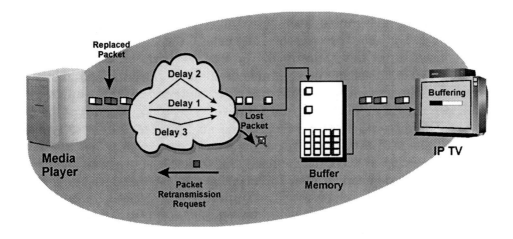

Figure 1.11, Packet Buffering

## Converting Packets to Television Service

IPTV data packets are converted back to television signals via gateways. Gateways may interconnect IPTV service to a television network (such as a hotel television system) or they may convert the signals directly to a television signal format (such as a NTSC or PAL analog television signals).

Gateways (Adapters) Connect the Internet to Standard Televisions

A television gateway (an adapter box) is a communications device or assembly that transforms audio and video that is received from a television media server (IPTV signal source) into a format that can be used by a viewer or different network. A television gateway usually has more intelligence (processing function) than a data network bridge as it can select the video and voice compression coders and adjust the protocols and timing between two dissimilar computer systems or IPTV networks.

Figure 1.12 shows how a media gateway connects a television channel to a data network (such as a broadband Internet connection). This example shows that the gateway must convert audio, video and control signals into a format that can be sent through the Internet. While there is one communication channel from the gateway to the end viewer, the communication channel carries multiple media channels including video, audio, and control information. The gateway first converts video and audio signals into digital form. These digital signals are then analyzed and compressed by a coding processor. Because end users may have viewers that have different types of coders (such as MPEG and AAC), the media gateway usually has available several different types of coding devices. This example shows that the media gateway receives requests to view information (the user or network sends a message to the media gateway). The gateway may have a database (or access to a database) that helps it determine authorized users and the addresses to send IPTV signals.

## Managing the Television Connections

Middleware software controls the setup, connection, feature operation, and disconnection of television channels connected through the data network. Middleware controls the media servers and gateways that provide media to the viewing devices. IPTV systems manage the downloading or streaming of IPTV signals to the consumer and may manage the selection (switching) of the media source.

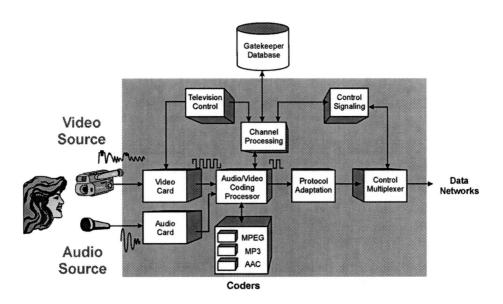

Figure 1.12, IPTV Gateways

Downloading

Downloading is the process of transferring a program or a data from a computer server to another computer. Download commonly refers to retrieving complete files from a web site server to another computer. Downloading movies requires the storage of the entire file on a hard disk or other type of memory storage for playback at a later time.

One of the key advantages of downloading is that the data transmission rate can vary and operate at any speed. If the user is willing to wait long enough, the entire file can be downloaded for future playback on any connection. Key disadvantages to downloading are the need to wait until the entire file is transferred before being able to view and the need to have enough storage room to hold the entire video (about 4 GB for a 2 hour movie).

Figure 1.13 shows how to download movies through the Internet. This diagram shows how the web server must transfer the entire media file to the media player before viewing can begin.

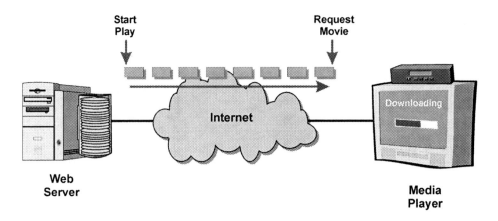

Figure 1.13, Downloading Movies through the Internet

Streaming

Streaming is a process that provides a continuous stream of information that is commonly used for the delivery of audio and video content with minimal delay (e.g. real-time). Streaming signals are usually compressed and error-protected to allow the receiver to buffer, decompress, and time sequence information before it is displayed in its original format.

Figure 1.14 shows how to stream movies through an IP data network. This diagram shows that streaming allows the media player to start displaying the video before the entire contents of the file have been transferred. This diagram also shows that the streaming process usually has some form of feedback that allows the viewer or receiving device to control the streaming session and to provide feedback to the media server about the quality of the connection. This allows the media server to take action (such as increase or decrease compression and data transmission rate) if the connection is degraded or improved.

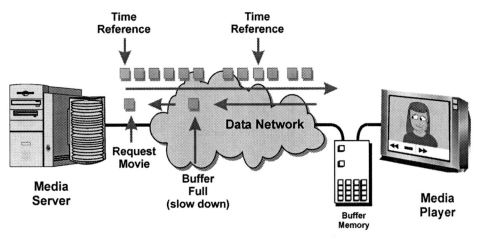

Figure 1.14, IPTV Streaming

Switching (Connecting) Media Channels

IPTV systems may set up connections directly between IPTVs or IP set top boxes and media servers or it may use a video switching system to connect the viewer to one of several available media sources. When the media connection is set up directly between the media server and the viewer, this is known as soft switching.

Figure 1.15 shows how a basic IPTV system can be used to allow a viewer to have access to many different media sources. This diagram shows how a standard television is connected to a set top box (STB) that converts IP video into standard television signals. The STB is the gateway to an IP video switching system. This example shows that the switched video service (SVS) system allows the user to connect to various types of television media sources including broadcast network channels, subscription services, and movies on demand. When the user desires to access these media sources, the control commands (usually entered by the user by a television remote control) are sent to the SVS and the SVS determines which media source the user desires to connect to. This diagram shows that the user only needs one video channel to the SVS to have access to virtually an unlimited number of video sources.

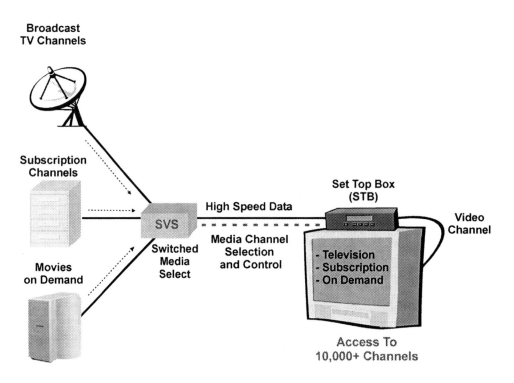

Figure 1.15, IPTV Connection

## Multiple IPTVs per Home

Each household may have several users that desire to watch different programs. This requires the bandwidth to be shared with each individual IPTV.

Households may have a combination of several multimedia computers, set top boxes or IPTVs in each home. When viewers are watching television channels (different channels), the bandwidth of each IPTV signal must be added.

Figure 1.16 shows how much data transfer rate it can take to provide for multiple IPTV users in a single building. This diagram shows 3 IPTVs that require 1.8 Mbps to 3.8 Mbps to receive an IPTV channel. This means the broadband modem must be capable of providing 5.4 Mbps to 11.4 Mbps to allow up to 3 IPTVs to operate in the same home or building.

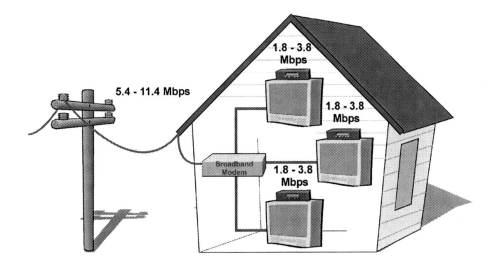

Figure 1.16, IPTV Multiple Users

## Transmission

IPTV channel transmission is the process of transferring the television media from a media server or television gateway to an end customer. IPTV channel transmission may be exclusively sent directly to a specific viewer (unicast) or it may be copied and sent to multiple viewers at the same time (multicast)

Unicast

Unicast transmission is the delivery of data to only one client within a network. Unicast transmission is typically used to describe a streaming connection from a server to a single client.

Unicast service is relatively simple to implement. Each user is given the same address to connect to when they desire to access that media (such as an IPTV channel). The use of unicast transmission is not efficient when many users are receiving the same information at the same time because a separate connection for each user must be maintained. If the same media source is accessed by hundreds or thousands of users, the bandwidth to that media server will need to be hundreds or thousands of times larger than the bandwidth required for each user.

Figure 1.17 shows how IPTV systems can deliver the same program to several users using unicast (one-to-one) channels. This example shows that each viewer is connected directly to the media server. Because each viewer is receiving 3 Mbps, the media server must have a connection that can provide 9 Mbps (3 Mbps x 3 viewers).

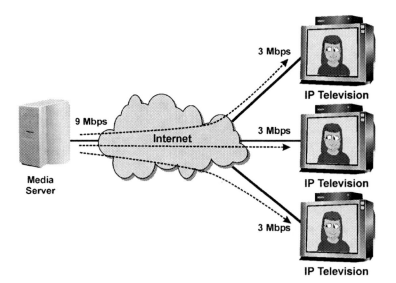

Figure 1.17, IPTV Unicast Transmission

Multicast

Multicast transmission is a one-to-many media delivery process that sends a single message or information transmission that contains an address (code) that is designated to allow multiple distribution nodes in a network (e.g. routers) to receive and retransmit the same signal to multiple receivers. As a multicast signal travels through a communication network, it is copied at nodes within the network for distribution to other nodes within the network. Multicast systems form distribution trees of information. Nodes (e.g. routers) that copy the information form the branches of the tree.

The use of multicast transmission can be much more efficient when the same information is sent to many users at the same time. The implementation of multicast systems is generally more complex than unicast systems as more control is required to add and remove members of multicast groups. Multicast recipients generally submit requests to a nearby node within a multicast network to join as part of an active multicast session.

For multicast systems to operate, nodes (routers) within the network must be capable of multicast sessions. Because of the complexity and cost issues, many Internet routers do not implement multicast transmission. If the multicast network is controlled by a single company (such as a DSL or cable modem data service provider), all the nodes within the network can be setup and controlled for multicast transmission.

Figure 1.18 shows how an IPTV system can distribute information through a switched telephone network. This example shows that end users who are watching a movie that is initially supplied by media center that is located some distance and several switches away from end users (movie watchers). When the first movie watcher requests the movie, it is requested from the telephone end office. The telephone end office determines that the movie is not available in its video storage system and the end office switch requests the movie from the interconnection switch. The interconnection switch also determines the movie is not available in its video storage system and the movie is requested from the distant media source. When the movie is transferred from the media center to the end customer, the interconnecting switches may make a copy for future distribution to other users. This program distribution process reduces the interconnection requirements between the switching distribution systems.

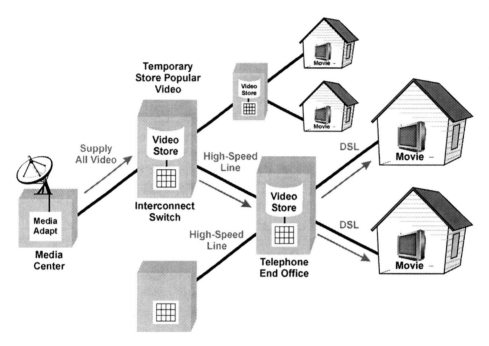

Figure 1.18, IPTV Multicast Transmission

## Viewing IPTV

IPTV media can be viewed on a multimedia computer, standard television using an adapter, or on a dedicated IPTV.

## Multimedia Computer

A multimedia computer is a data processing device that is capable of using and processing multiple forms of media such as audio, data and video. Because many computers are already multimedia and Internet ready, it is often possible to use a multimedia computer to watch IPTV through the addition or use of media player software. The media player must be able to find and connect to IPTV media servers, process compressed media signals, maintain a connection, and process television control features.

Control of IPTV on a multimedia computer can be performed by the keyboard, mouse, or external telephone accessory device (such as a remote control) that can be connected to the computer through an adapter (such as an infrared receiver). The media player software controls the sound card, accessories (such as a handset), and manages the call connection.

IPTV signals may be able to be displayed on a multimedia provided it has enough processing power (processing speed) and the necessary media player protocols and signal decompression coders. IPTV signals contain compressed audio and video along with control protocols. These signals must be received, decoded and processed. The processing power of the computer may be a limitation for receiving and displaying IPTV signals. This may become more apparent when IPTV is taken from its small format to full screen video format. Full screen display requires the processor to not only decode the images but also to scale the images to the full screen display size. This may result in pixilation (jittery squares) or error boxes. Using a video accelerator card that has MPEG decoding capability may decrease the burden of processing video signals.

A media player must also have compatible control protocols. Just because the media player can receive and decode digital video and digital audio signals, the control protocols (e.g. commands for start, stop, and play) may be in a protocol language that the media player cannot understand.

## IP Set Top Boxes (IP STB)

An IP Set Top box is an electronic device that adapts IP television data into a format that is accessible by the end user. The output of an IP set top box can be a television RF channel (e.g. channel 3), video and audio signals or digital video signals. IP set top boxes are commonly located in a customer's home to allow the reception of IP video signals on a television or computer.

An IP STB is basically a dedicated mini computer which contains the necessary software and hardware to convert and control IPTV signals. IP STBs must convert digital broadband media channels into the television (audio and video signals) and decode and create the necessary control signals that pass between the television and the IPTV system.

## IP Televisions

IP televisions are display devices that are specifically designed to receive and decode television channels from IP networks without the need for adapter boxes or media gateways. IP televisions contain embedded software that allows them to initiate and receive television through IP networks using multimedia session protocols such as SIP.

An IP television has a data connection instead of a television tuner. IP televisions also include the necessary software and hardware to convert and control IPTV signals into a format that can be displayed on the IP television (e.g. picture tube or plasma display).

## Mobile Video Telephones

Mobile telephones with multimedia capabilities may be able to watch television channels. Mobile telephones have limited information processing power, limited displays, and may have restricted access to Internet services.

Multimedia mobile telephones contain embedded software that allows them to initiate and receive multimedia communication sessions through the

Internet. Because of the limited bandwidth and higher cost of bandwidth for mobile telephones, mobile telephone media players in mobile telephones may use compression and protocols that are more efficient than are used by standard IPTV systems. To increase the efficiency, mobile telephone data sessions may be connected through gateways that compress media signals and convert standard control protocols to more efficient and robust control protocols. This may cause some incompatibilities or unexpected operation for mobile television systems.

# Chapter 2

# IPTV Services and Features

IPTV services are the providing of multimedia services (e.g. television) to customers by a common carrier, administration, or private operating agency using voice, data, and/or video technologies. In addition to providing the basic television services and features, IPTV can provide advanced features and services that are not possible with traditional broadcast television systems.

## Television Programming

Television broadcast services are the transmission of television program material (typically video combined with audio) that may be paid for by the viewer and/or by advertising. Television programming may be live, scheduled or on-demand programming. Live television broadcasting is the transmission of video and audio to a geographic area or distribution network in real time or near-real time (delayed up to a few seconds). Scheduled programming is the providing of television programs in a pre-selected time sequence. On demand programming is providing or making available programs that users can interactively request and receive.

IPTV television programming sources include traditional broadcast sources such as television networks, syndicates and local stations along with new content sources such as international (global) programs, sponsored programs, community content and personal media channels.

## Enterprise (Company) Television

Enterprise programming is media that is created and managed for viewing by a company or for visitors it authorizes to view its programming content. Company television may be produced for the public and/or for internal communication purposes. Public company television channels may provide information about products, services or applications of the products or services that are of interest to the public. Internal ("in-house") company television programs may be used to provide employees with educational and company specific information (such as the location of a company meeting or party). Employees, vendors or others who are provided with access may distribute company television programs to monitors within company buildings or for distribution to multimedia computers that are only accessible by company employees.

## Gaming

Gaming is an experience or actions of a person that are taken on a skill testing or entertainment application with the objective of winning or achieving a measurable level of success. Gaming services provided by IPTV systems may include game program distribution (downloading games), online gaming service, multi-user network gaming or gambling.

## Security Monitoring

Security systems are monitoring and alerting systems that are configured to provide surveillance and information recording for protection from burglary, fire, water hazard, and other types of losses. Video surveillance is the capturing of video for the observation of an area or location at another location.

Traditional (legacy) security systems use proprietary sensing and transmission equipment, have limited control processing capabilities, and have interconnections that are limited to local geographic areas. The use of IPTV systems connected through standard data networks allows for the sending of media (such as digital video), powerful security system processing in a server, and wide area connectivity (such as through the Internet).

IPTV access devices (such as set top boxes) may contain connection points (such as USB connections) that allow for digital video signals to be sent through the IPTV system to a monitoring system. These monitoring systems can be owned an operated by the IPTV system operator or they can be monitoring stations operated and managed by other companies (such as police station central monitoring facilities).

Figure 2.1 shows how a variety of security accessories can be integrated into an IPTV based communication system. This example shows how a police station can monitor multiple locations (several banks) through the addition of digital video and alarm connections. This example shows that when a trigger alarm occurs at a bank (such as when a bank teller presses a silent alarm button), the police can immediately see what is occurring at the bank in real-time. Because the images are already in digital format, it may be possible to send these pictures to police cars in the local area to help identify the bank robbers.

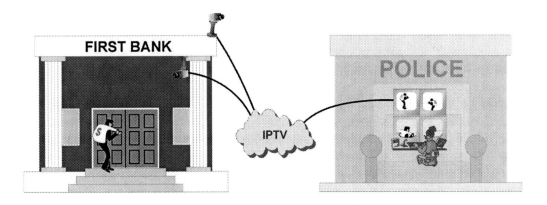

Figure 2.1., IPTV Security Monitoring

## Advertising

Advertising is the communication of a message or media content to one or more potential customers. One of the most complicated areas for IPTV can be the management of advertising services. Advertising management is the process of creating, presenting, managing, purchasing and reporting of advertising programs. Because advertising services on IPTV systems can range from broadcast advertising (to all people in a geographic area) to customized addressable advertising (custom ads for specific viewers), advertising management can be a complex but yet a very profitable process.

An advertising program typically begins by setting up advertising campaigns. Advertising campaigns define the marketing activities such as the specific advertising messages that will be sent to customers who are classified into certain categories (target market segments) about products, services, and options offered by a company.

Advertising messages may be in the form of interstitial, mixed media or interactive media. An interstitial ad is an advertising message that is inserted "in between" program segments. Interstitial ads can also be pop-ups (when selecting a new channel) and pop-downs (when exiting a selected program). Mixed media advertising is the combining of advertising media along with other video and text graphics on a television or video monitor. Interactive advertising is the process of allowing a user to select or interact with an advertising message.

Figure 2.2 shows how IPTV advertising messages can be in the form of interstitial broadcast messages, mixed media messages or interactive ads. In example A, a network operator provides a program with advertising messages already inserted (interstitial) into the program. Example B shows how an advertising message may be overlapped or merged into the underlying television program. Example C shows how an advertising message my change based on the selections of the viewer.

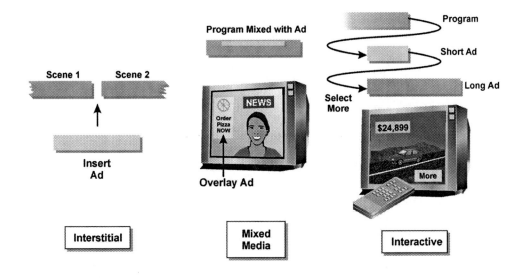

Figure 2.2, Types of IPTV Advertising Messages

## Television Commerce (T-Commerce)

Television commerce (t-commerce) is a shopping medium that uses a television network to present products and process orders. The processes that used in t-commerce include advanced product offering catalogs (video catalogs), order processing, exchanging of order information between companies in near real-time and the ability to offer multiple forms of payments that may be collected by different companies. Key issues for IPTV t-commerce billing include transferring accounting records through multiple systems that transfer between multiple companies that allow for presentation, processing and payment of orders.

Figure 2.3 shows how a television program can use mixed media to provide product offers to qualified consumers at specific times in a display location that is noticeable but not intrusive. This picture shows that during a news program, the viewer is presented with a pizza icon from a local pizza restaurant. This example shows that when the user selects the icon, a small window appears with the pizza offer details.

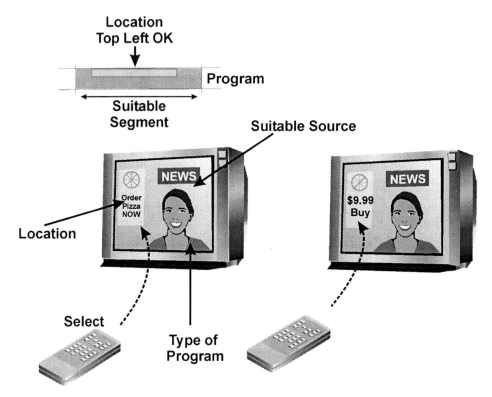

Figure 2.3, Television Commerce

Vendors of products that are sold on IPTV systems may be allowed to directly interact with their product offers as they know their customers better than IPTV service providers. This means that t-commerce systems will likely offer management portals. Offer management portals will allow the vendor to add new products, configure their presentation options for the product (e.g. mixed media) and define the product or service offers for specific market segments.

# IPTV Content

IPTV content is a mixture of rich media programming that includes television and other forms of content. Some of the common forms of IPTV content include network, syndicated, local, sponsored, community and personal content.

## Network Programming

Network programming is the selection of shows and programs that are offered by a television network provider. An example of a network program is a television series that is developed and owned by a television network. These programs are distributed to network affiliates (local broadcasters) along with advertising commercials that is sold by the network and/or local broadcaster.

## Syndicated Programming

A syndicated program is a media session (such as a television show) that is distributed by an organization to more than one broadcasting company. An example of a syndicated program is a television show that is developed independent of the networks and is made available to multiple television broadcasting companies.

## Local Programming

Local programming is the selection of shows and programs that are offered by a local television network provider. An example of a local program is a news program that is created and broadcasted by a local broadcaster.

## Sponsored Programming

Sponsored content is specialty programs or media that are paid for by people or companies who are looking to promote a specific solution or to develop a mailing list of the viewers who watch the content (similar to sponsored web seminars).

## International Programming

International programming is the selection of shows and programs that are offered by sources outside the country that is offering the television service. An example of a global program is a news program that is created and broadcasted from another country.

## Community Content

Community content programming is media that is created and managed by members of a community or a group that can be viewed by others who are interested in community content. Examples of community content include school, sports and local events that members of a community have an interest in. Community members are commonly interested in assisting in the creation, management and delivery of the community content with or without direct compensation.

## Personal Media

Personal content is media or data that is created and managed by an individual. The individual owner or user of personal content may establish a hierarchy of viewing rights that can vary between family, close friends or lists of specific individuals or channel types. Examples of personal content include pictures of friends, video clips at events and creative images.

IPTV systems offer cost effective access to any type of program including programs that have a very high initial interest (long tail content) and programs that have a low interest level for an extended period of time (flat tail content).

## Long Tail Content

Long tail content is programs or media that is initially viewed or desired by a large group of people and then has a much lower interest and viewing level that occurs over a relatively long period of time. Long tail typically refers to the statistical distribution of valuable products or content where a majority of usage or distribution occurs at the beginning of a process or product offering.

## Flat Tail Content

Flat tail content is program media that has a relatively consistent viewing demand over a period of time. Flat tail content is programs such as educational or personal development programs that provide enriching information to viewers when they need that information. As time passes, additional people begin to need this information which results in the continuous demand for flat tail content. An example of flat tail content is an instructional painting lesson (e.g. the "Painting in Watercolors") or a home repair show (e.g. "How to Install Tile").

# IPTV System Features

IPTV system features include instant activation, real time billing, interactive program guides, advanced channel selection options, interactive television control, anywhere television service, global television channels, personal media channels, addressable advertising, ad telescoping, everything on demand, ad bidding, t-commerce order processing and user profiling.

## Instant Activation

Instant activation is a process that allows users to obtain service immediately after applying for service. Instant activation may use the same data connection for requesting and activating communication services as for the transfer of the required services and features.

Figure 2.4 shows how it is possible for a user or company system administrator to instantly activate a new IPTV service. In this example, the ITVSP has created a web access page that allows the user to self activate themselves. After the user has provided the necessary information, such as a billing address and method of payment, account identification codes can be provided manually to the user or they may be automatically entered into the IPTV viewer. The user can then select feature preferences such as preferred television channels and viewer profile.

Figure 2.4, IPTV Instant Line Activation

## Real Time Accounting and Billing

Real time accounting and billing is the process of gathering, rating, and displaying (posting) of account information either at the time of service request or within a short time afterwards (may be several minutes). IPTV service commonly allows for real time billing for tracking of IPTV services.

Figure 2.5 shows how IPTV service can provide real time accounting and billing records immediately after they are created (in real time). This example shows how the IPTV service provider keeps track of each channel (or premium channel used). It uses the television channel setup and termination information to adjust the accounting and billing information. In this example, these charges or usage amounts can be displayed immediately through on the IPTV viewer or through an Internet web page.

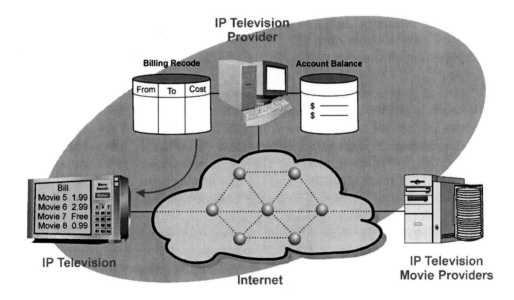

Figure 2.5, IPTV Real Time Accounting and Billing

## Channel Selection (Program Guides)

IPTV channel selection is the process of finding and connecting to an IPTV data address (IP address) so it can receive and decode a television or media channel.

It is possible for viewers to direct connect to IPTV channels if they know the URL or IP address (web link). Although the viewer may have the address or the URL of the IPTV channel (a media server), the viewer may not be authorized to connect to the channel at that address. Owners of IPTV media can restrict access to paying customers. Authorization codes are typically pre-established by viewers or companies that provide IPTV services to viewers.

IPTV service providers provide a selection screen or device that allows users to find and select IPTV channels. Because it is difficult for viewers to remember or organize URLs and IP addresses, channel selection screens usually have more descriptive information such as channel numbers, network names and show titles. While it is possible to have IPTV service that uses channel numbers that are identical to standard cable television systems, IPTV service providers (ITVSPs) offer many new ways to find and select television channels.

IPTV service providers usually provide an electronic programming guide (EPG) as an interface (portal) that allows a customer to preview and select from possible lists of available content media. EPGs can vary from simple program selection to interactive filters that dynamically allow the user to filter through program guides by theme, time period, or other criteria. Viewers are also able to connect to IPTV channels through the use of a web link on web pages or through a link that is sent (embedded) in emails.

Figure 2.6 shows some of the different ways a user can find and select IPTV channels. While it is possible for IPTV systems to use channel numbers for the selection of IPTV channels, this example shows that there are several new more effective ways to search and select channels. The user can search for channels by favorites, country, actor name, show title, network provider and category. The user can also select from the channel numbers offered by their IPTV provider.

| Channel Selection | | | 2 |
| --- | --- | --- | --- |

Figure 2.6, IPTV Channel Selection

## Interactive Television (iTV)

Interactive television is the providing of video services that allows for the user to control part or all of the viewing experience. Interactive television has three basic types: "pay-per-view" involving programs that are independently billed, "near video-on-demand" (NVOD) with groupings of a single film starting at staggered times, and "video-on-demand" (VOD), enabling request for a particular film to start at the exact time of choice. Interactive television offers interactive advertising, home shopping, home banking, e-mail, Internet access, and games.

## Anywhere Television Service (TV Roaming)

Television extensions are viewing devices that are connected to a television distribution system. These connections may be shared (several televisions on the same line) or they may be independently controlled (such as in a private television system). Traditionally, television extensions have a fixed wire or a connection line. This allows a television viewing device to either share (directly connect to) another communication line or to allow it to independently connect it to a switching point (such as a private company Television system).

When an IPTV viewer is first connected (plugged-in) to a data connection, it requests the assignment of a temporary Internet address from the data network. It uses this Internet address to register with the ITVSP after it has been connected to the Internet. Because the ITVSP always knows the current Internet address that is assigned to the IPTV each time it has been connected to the Internet, this allows IPTVs to operate at any connection point that is willing to provide it broadband access to the Internet. In essence, this allows an IPTV to operate like a television extension that can be plugged in anywhere in the world.

What makes this so interesting is that IPTVs can be taken anywhere and they will continue to operate as if there were no changes. Suppose a person in Singapore subscribes to local television services in Singapore. If this person takes their IPTV viewer (possibly a laptop) to a hotel in New York that has a broadband connection, when they plug the IPTV viewer into the Internet in New York, it will operate just as it was in their home in Singapore. It does this because the IPTV registers with the IPTV service provider when it is plugged into the Internet. The Internet address of each IPTV is dynamically assigned each time they are turned on and the IPTV system keeps track of these addresses. It doesn't really matter where the IPTV is plugged in.

Figure 2.7 shows how Internet Televisions can be used anywhere in the world where they can be connected to an Internet connection. In this example, a user subscribes to Internet television service from an IPTV service provider in Singapore. The IPTV service provider manages user accounts and sets up connections between users and media sources. This diagram shows that an Internet television user can obtain television service at different locations such as at an airport, at home, in the office, or in a hotel. As long as the Internet television user has a valid Internet television account, they can use different types of viewers such as multimedia computers (laptops), standard televisions with adapter boxes and IPTVs.

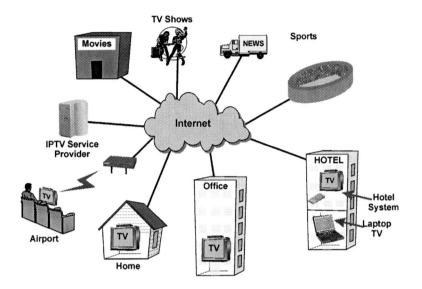

Figure 2.7, Internet Television Extensions go Anywhere in the World

## Global Television Channels

Global television channels are TV channels that can be viewed anywhere in the world. IPTV channels that are delivered through the Internet typically can be viewed at any location in the world that offers broadband data access.

The ability of IPTV systems to provide video service to outside their local (often regulated) areas allows for new competition. The typical cost for viewing global television channels is the content media access costs (such as a fee for a movie) and the broadband data access cost (a monthly broadband access charge).

Figure 2.8 shows how IPTV allows a person to view television channels in other cities, states, or countries. In this example, a viewer from the United Kingdom is traveling to the United States and he desires to watch a soccer game in Australia. This diagram shows that the viewer connects his multimedia laptop to a broadband connection and logs into his Internet television service provider with his access codes. He then selects the channel source and quality of image he desires (the rates increase for higher quality access). The viewers ITVSP provider is located in the UK. When the ITVSP receives and authorizes the request, it sends an authorization message to a media gateway in Australia that is converting the video from the sports event to a data format that can be sent through the Internet. The media gateway then sends the information directly to the viewer through the Internet.

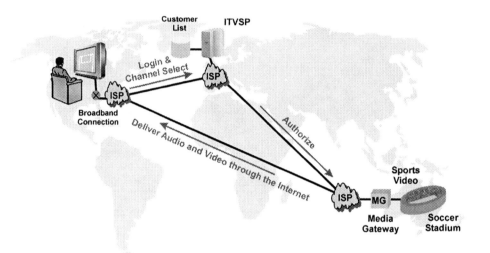

Figure 2.8, Using Internet Television to view International Television Channels

## Personal Media Channels (PMC)

A personal media channel (PMC) is a communication service that allows a media user (e.g. viewer) to select and view media (typically video or music) from a variety of media sources.

An example of how a PMC may be used for IPTV is the control and distribution of mixed media (such as digital pictures and digital videos) through a personal television channel to friends and family members. An IPTV customer may be assigned a personal television channel. The user can upload media to their personal media channels and allow friends and family to access pictures and videos of family members and gatherings via their IPTVs.

Figure 2.9 shows how personal media channels allow other viewers to be given access to specific types of personal media on their IPTV. This example shows how an IPTV user "Bob" is uploading pictures from a party to his personal media channel 9987. People on his list of viewers who also have IPTV service can select Bobs television channel 9987 and see the pictures. This example shows that the personal media channel can have restrictions on who can view the personal media channel.

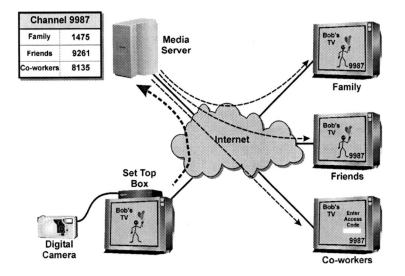

Figure 2.9, Personal Media Channels

## Addressable Advertising

Addressable advertising is the communication of a message or media content to a specific device or customer based on their address. The address of the customer may be obtained by searching viewer profiles to determine if the advertising message is appropriate for the recipient. The use of addressable advertising allows for rapid and direct measurement of the effectiveness of advertising campaigns.

A key aspect of addressable advertising is the validation of the viewer. IPTV systems may ask (prompt) the viewer to select their name from a list of registered users in the home when the IPTV is turned on. Because of the advanced features offered by IPTV such as incoming calls/emails and programming guides that remember favorite channels, viewers will typically want to select their programming name. Because the programming name has a profile (preferences), advertising messages can be selected that best match the profile.

The potential revenue for addressable advertising messages that are sent to viewers with specific profiles can be 10 to 100 times higher than the revenue for broadcasting an ad to a general audience. The ability to send ads to a specific number of viewers allows advertisers to set specific budgets for addressable advertising. It also allows the advertiser to test a number of different ads in the same geographic area at the same time.

Figure 2.10 shows how addressable advertising can be used to better match advertising messages to the wants and needs of viewers. This diagram shows that a media program (such as a television show or movie) is being

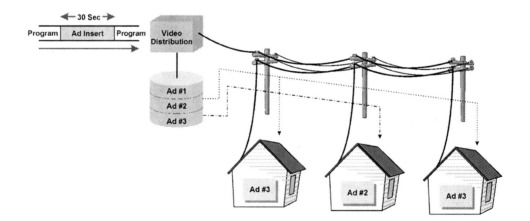

Figure 2.10, Addressable Advertising

sent to 3 homes where the televisions in each home have a unique address. When the time for a 30 second commercial occurs, a separate advertising message is sent to each one of the viewers based on the address of the television. This allows each viewer to receive advertising messages that are better targeted to their needs and desires.

## Ad Telescoping

Telescoping advertisements are extended advertising messages (selected or automatically expanded) from a smaller and/or shorter version of an ad to a larger and/or longer version of an ad. Ad telescoping allows the viewer to immediately obtain more information about a product or service by selecting an interactive option on the advertising message.

The viewer may be provided with options of immediately viewing the expanded ad or bookmarking an ad for later viewing. If the ad is expanded, their current viewing point in the television program may be stored so they can return to the exact point they left when they selected to view the expanded ad (time shifting).

Figure 2.11 shows how ads can be expanded using ad telescoping. This diagram shows that a viewer is presented with an ad that can be expanded to provide more information. If the viewer selects the more button, the channel source is redirected to a longer (expanded) advertising message.

## Everything on Demand (EoD)

Everything on demand is a service that provides end users to interactively request and receives media services of any type. These media services may come from previously stored media (entertainment movies or education videos) or they may have a live connection (news events in real time). Everything on demand services include video on demand (VOD), near video on demand (NVOD) and time shifted programming.

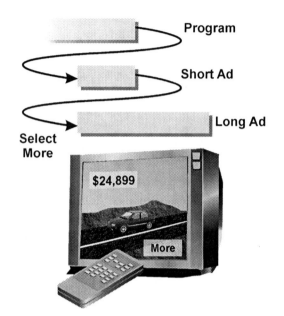

Figure 2.11, Ad Telescoping

Video on demand is a service that provides end users to interactively request and receive video services. These video services are from previously stored media (entertainment movies or education videos) or have a live connection (news events in real time).

Near video on demand is a video delivery service that allows a customer to select from a limited number of broadcast video channels when they are broadcast. NVOD channels have pre-designated schedule times and are used for pay-per-view services.

EoD can be enhanced by advanced electronic programming guides that maintain a history of previously viewed television or media shows. This allows the viewer to scan through a list of programs that they have not previously viewed.

Figure 2.12 shows how IPTV can allow a viewer to request control of the presentation of television programs on demand. This diagram shows that a television on demand viewer can browse through available television channels. In this example, this IPTV service provider informs the viewer of which programs they have already viewed and the length of time each program will run. When the user selects a potential program to view, a short description of that program is shown at the bottom of the screen along with the cost for viewing that particular program.

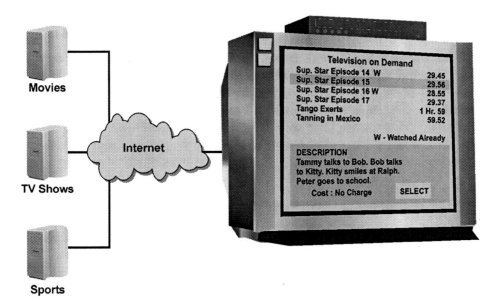

Figure 2.12, Everything on Demand

## Ad Bidding

Ad bidding is the process of selection of threshold amounts that may be paid for the insertion of advertising messages. Ad bidding requires a bid management process that can monitor and adjust the bid amounts for the requested insertion and placement of ad messages. An example of ad bidding is the paying for ad listings on search engines such as Google or Yahoo.

Bidding for IPTV systems is likely to occur for particular segment types. For example, it may be possible to separately bid for different age groups. Because IPTV systems can interact with the viewer, it may be possible to determine which segment the viewer belongs to. For example, if each viewer in a household is provided with a login choice so they can customize their interface (TV screen saver) and access their preferred channel list (MTV compared to Discovery Channel), it is likely that the actual viewer and their characteristics can be determined.

Figure 2.13 shows how IPTV ad bidding may work for selling cars. This diagram shows that bidding for advertising messages may occur for particular age groups, income ranges, program types, geographic regions. This example shows that IPTV advertisers may bid for ads that may appear on a variety of programs throughout various geographic regions. The advertiser sets the maximum bid they are willing to offer and a maximum number of impressions that may be selected to ensure advertising budgets can be maintained. This example shows that the advertiser may also be able to select if the same ad should be sent to the same person more than one time.

| Ad Name | Age Group | Income Range | Program Type | Regions | Ad Repeats | Bid per Impression | Max Impressions |
|---------|-----------|--------------|--------------|---------|------------|--------------------|-----------------|
| Utility Vehicle | 25-39 | 70k+ | Sports | Nationwide | Yes | 0.10 | 10,000 |
| Status Auto | 40-54 | Any | Entertainment | Nationwide | No | 0.18 | 20,000 |
| Luxury Car | 55-69 | 70k+ | Travel | Florida | Yes | 0.07 | 10,000 |

Figure 2.13, Ad Bidding

Measuring the performance and success of an advertising program can be accomplished through the use of existing and new types of marketing measurements including ad impressions, ad selections, ad expansions and ad compressions.

To help companies determine the success of their advertising campaigns, advertising reports are tables, graphs or images that may be provided to represent specific aspects of advertising campaigns or the information or data that is created from advertising campaign. IPTV advertising reports may include the number of ad impressions per segment, number of click through selections, the number of ad expansions and the number of ad compressions.

An ad impression is the presentation of an advertising message or image to a media viewer. Ad selections are the clicking or indication that a button or attribute on an advertising message has been selected. Ad selections are indicated by the click through rate (usually in percentage form), which is a ratio of how many selections (red button) or clicks (mouse selections) an advertising message or item within the ad message receives from visitors compared to the number of times the advertising message is displayed. An example of click through rate for an ad button that is clicked 5 times out of 100 displays to visitors is 5%.

IPTV offers the opportunity for interactive advertising, which allows a user to select or interact with an advertising message. This interaction may result in a redirecting of the source of an advertising message to play a longer more informative version of the ad (expanded ads). Viewers may also be able to end an advertising message to return to their media program (compressed ads).

Figure 2.14 shows a sample report that may be generated for IPTV interactive advertising. This example shows that an advertiser has selected to advertise to three age groups.

| Age Group | Impressions | Click Through |
|---|---|---|
| 5 through 14 | 0 | 0 |
| 15 through 24 | 0 | 0 |
| 25 through 39 | 416 | 4 |
| 40 through 54 | 347 | 11 |
| 55 through 69 | 287 | 1 |
| over 70 | 0 | 0 |

FFigure 2.14, IPTV Advertising Reporting

## T-Commerce Order Processing

Television commerce order processing is the steps involved in selecting the products or service from a television catalog or advertising and agreeing to the terms that are required for a person or company to obtain products or services.

For t-commerce systems, after a viewer has selected a product offer, order processing occurs. Order processing is the defining of terms that are agreeable to the viewer for the acquisition of a product or service. Selected products are placed in a shopping cart for the particular user. Shopping carts are the electronic containers that hold online store items while the user is shopping. The online shopper is typically allowed to view and change items in their shopping cart until they purchase. Once they have completed the purchase, the items are removed from their shopping cart until they start shopping again.

It is important that the t-commerce system identify the particular user as there can be several users in a household that share an IPTV and each may have orders in progress with a variety of vendors.

Another important part of the t-commerce system is the fulfillment process. Fulfillment is the process of gathering the products and materials to complete an order and shipping the products or initiating the services that were ordered. Depending on the types of products and services offered, IPTV order fulfillment can range from the immediate delivery of media products (such as games or television programs) to the delivery of products over an extended period of time (such as an order of books that has a mix of available and future ship dates).

Customers will likely associate responsibility for fulfilling the order to the TV service provider. To reduce the cost of customer care and to avoid potential negative conflicts for unfulfilled products, t-commerce systems may include order tracking capability. Order tracking is the ability of a customer, company or other person who is involved with an order to gather information as to the status of the processing of the order.

Figure 2.15 shows a typical scenario of t-commerce order processing. This example shows that a viewer is presented with a product offer (a pizza). This offer is associated with an offer identification code to allow the user to select the offer and to be redirected to an order window. When the user completes the order, the order information is sent to the vendor (the pizza restaurant) where it is confirmed. This diagram shows that order status information may be provided from the vendor to the TV service provider and this information may be used to update the customer about the status of the order (pizza cooking). When the order is complete, the vendor provides information to the TV service provider that the order has been filled to allow the order record to be marked as completed.

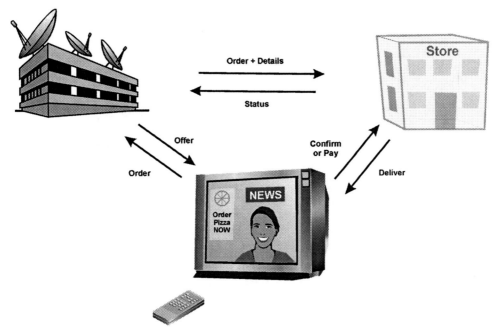

Figure 2.15, Television Order Processing

T-commerce orders can have a variety of payment methods that may need to be recorded in near real time to ensure the vendor and the t-service provider receive payment when the provide products or services.

Payment processing is the tasks and functions that are used to collect payments from the buyer of products and services. Payment systems may involve the use of money instruments, credit memos, coupons, or other form of compensation used to pay for one or more order invoices. T-commerce payment options include payment on the television bill, direct payment collection by the vendor, bill to $3^{rd}$ party or pay using other payment options. It is likely that t-commerce will offer a mixed set of payment options to the viewer.

Figure 2.16 shows how a vendor may receive a t-commerce order report. This example shows that a t-commerce vendor may receive payment from a t-commerce customer directly by cash or a credit card transaction, the customer may be able to place the order on their television bill or the customer may use a $3^{rd}$ party such as Paypal to pay for the transaction.

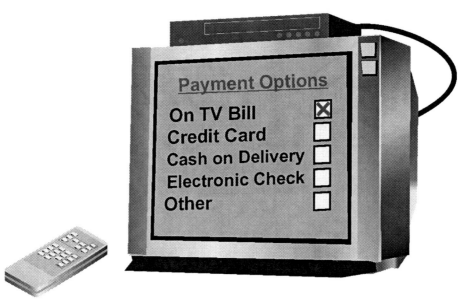

Figure 2.16, T-Commerce Payment Options

To allow multiple companies to process orders with multiple TV service providers, a standardized billing communication system is necessary. This billing system will need to transfer a variety of event information including order details, order status and payment information.

## User Profiling

User profiling is the process of monitoring, measuring and analyzing usage characteristics of a user of a product or service. IPTV service offers the possibility for recording and using viewer information to better target services to users. An example of user profiling is the offering of movie service packages (such as 5 films for children at a discount price) that is based on the previous viewing habits (such as watching 5 children's movies in the past 2 weeks).

# Chapter 3

# Introduction to IP Video

IP video is the transfer of video information in IP packet data format. Transmission of IP video involves digitizing video, coding, addressing, transferring, receiving, decoding and converting (rendering) IP video data into its original video form.

Figure 3.1 shows how video can be sent via an IP transmission system. This diagram shows that an IP video system digitizes and reformats the original video, codes and/or compresses the data, adds IP address information to each packet, transfers the packets through a packet data network, recombines the packets and extracts the digitized audio, decodes the data and converts the digital audio back into its original video form.

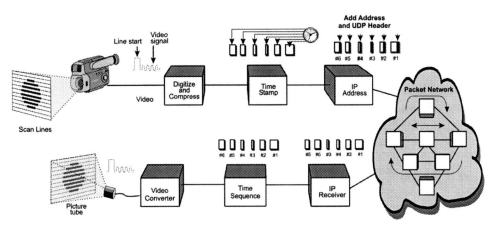

Figure 3.1, IP Video System

## Resolution

Resolution is the number of image elements (pixels) per unit of area. A display with a finer grid contains more pixels, and therefore has a higher resolution capable of reproducing more detail in an image.

Resolution is usually defined as the amount of resolvable detail in the horizontal and vertical directions in a picture. Horizontal resolution usually is expressed as the number of distinct pixels or vertical lines that can be seen in the picture. Vertical resolution is the amount of resolvable detail in the vertical direction in a picture. Vertical resolution usually is expressed as the number of distinct horizontal lines that can be seen.

A pixel is the smallest component in an image. Pixels can range in size and shape and are composed of color (possibly only black on white paper) and intensity. The number of pixels per unit of area is called the resolution and more pixels per unit area provide more detail in the image. Each pixel can be characterized by the amount of intensity and color variations (depth) it can have.

Pixel depth is the number of bits per pixel that is used to represent intensity and color. Pixel color depth is the number of bits per pixel that is used to represent color. For example, a color image that uses 8 bits for each color uses 24 bits per pixel. For a low color depth like 8 bits per pixel, a color palette may be used to map each value to one of a small number of more precisely represented colors, for example to one of 256 24-bit colors.

Figure 3.2 shows how the resolution of a video display is composed of horizontal and vertical resolution components. This diagram shows that the vertical resolution is described as lines per inch (lpi) and the horizontal resolution is described as dots (pixels) per inch (dpi).

## Frame Rates

Frame rate is the number of images (frames or fields) that are displayed to a movie viewer over a period of time. Frame rate is typically indicated in frames per second (fps).

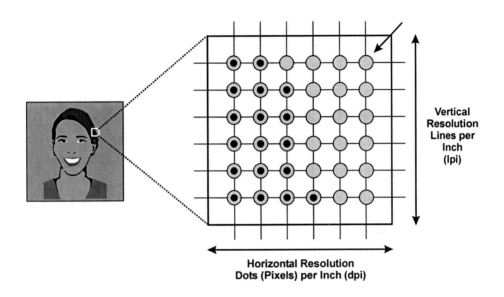

*Figure 3.2, Video Resolution*

In general, the higher the frame rate, the better the quality of the video image. When the frame rate is too low, flicker (fluctuations in the brightness of movie images) begins to occur. Flicker typically happens when the frame rate is below 24 frames per second. Some of the common frame rates for moving pictures are 24 fps for film, 25 fps for European video, 30 fps for North American video, 50 fps for European television and 60 fps for North American television.

Figure 3.3 shows the different types of frame rates and how lower frame rates can cause flicker in the viewing of moving pictures. This example shows that frame rates are the number of images that are sent over time (1 second). At the frame rate is reduced below approximately 24 frames per second (fps), the images appear to flicker. Increasing the frame rate results in increased bandwidth requirements. The common frame rate formats are 24 fps for film, 25 fps for European video, 30 fps for North American video, 50 fps for European television and 60 fps for North American television.

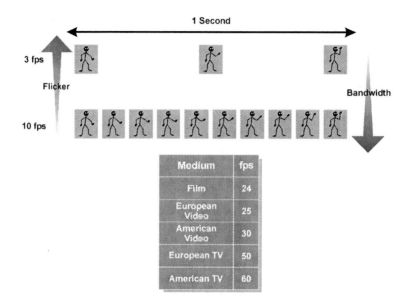

*Figure 3.3, Moving Picture Frame Rates*

## Aspect Ratio

Aspect ratio is the ratio of the number of items (such as pixels on a screen) as compared to their width and height. The aspect ratio determines the frame shape of an image. The aspect ratio of the NTSC (analog television) standard is 4:3 for conventional monitors such as home television sets and 16:9 for HDTV (widescreen).

Figure 3.4 shows how aspect ratio is the relationship between width and height expressed as width:height. This diagram shows that wide screen television has an aspect ratio of 16:9 and that standard television and computer monitors have an aspect ratio of 4:3.

**Aspect Ratio is (width) : (height)**

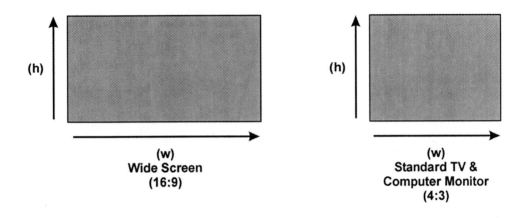

Figure 3.4, Aspect Ratio

## Letterbox

Letterbox is the method of displaying wide screen images on a standard TV receiver where the wide screen aspect ratio is much larger than the standard television or computer monitor typical aspect ratio of 4:3. This causes the display to appear within borders at the top and bottom of the image producing a horizontal box (the letterbox).

Figure 3.5 shows how the use of a letterbox allows an entire video image to be displayed on a screen that has an aspect ratio lower than the video image requires. This example shows that part of an image (of a boat) is lost on the left and right parts of the display. Using a letterbox, the image size is reduced so its width can fit within the length of the screen area. The result is part of the top and bottom areas of the screen area are blanked out (black) resulting in the formation of a box (a letterbox).

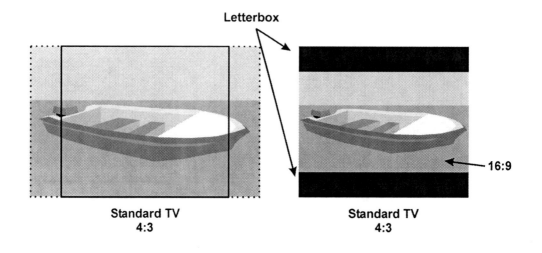

Figure 3.5, Letterbox

## Analog Video

Analog video is the representation of a series of multiple images (video) through the use of a rapidly changing signal (analog). This analog signal indicates the luminance and color information within the video signal.

Sending a video picture involves the creation and transfer of a sequence of individual still pictures called frames. Each frame is divided into horizontal and vertical lines. To create a single frame picture on a television set, the frame is drawn line by line. The process of drawing these lines on the screen is called scanning. The frames are drawn to the screen in two separate scans. The first scan draws half of the picture and the second scan draws in between the lines of the first scan. This scanning method is called interlacing. Each line is divided into pixels that are the smallest possible parts of the picture. The number of pixels that can be displayed determines the resolution (quality) of the video signal. The video signal television picture into three parts: the picture brightness (luminance), the color (chrominance) and the audio.

## Component Video

Component video consists of three separate primary color signals: red, green, and blue (RGB). The combination of component video can produce any color and intensity of picture information.

## Composite Video

Composite video is a single electrical signal that contains luminance, color, and synchronization information. Composite video signals are created by combining several analog signals. Examples of composite video formats include Y, U and V. NTSC, PAL, and SECAM.

The Y intensity (brightness) component includes can be used as a black and white (monochrome) video. The Y component contains the synchronization signal (sync pulse) that coordinates the picture tube scanning (image creation) process. The color components are represented by the two signals U and V. U is the difference between Blue and Y (intensity) and V is the different between Red and Y. The U and V are mixed with different phases (orthogonal) of a color carrier signal and combined to form a chrominance signal.

Figure 3.6 shows how a composite video signal is combines video intensity (Y) and color signals (U and V) to produce a composite video signal. This example shows that an intensity only (black and white) signal is combined with color component signals. The color component signals are added out of phase relative to each other.

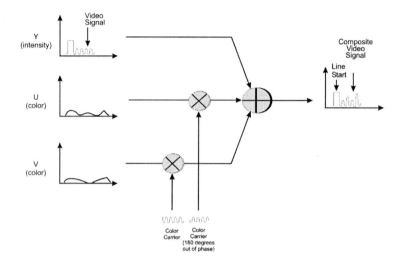

Figure 3.6, Composite Video

## Separate Video (S-Video)

S-video is a set of electrical signals that represent luminance and color information. S-video is a type of component video as the intensity and color components are sent as separate signals. This allows the bandwidth of each component video part to be larger than for the signals combined in composite video and it eliminates the cross talk distortion that occurs when multiple signals are combined. This offers the potential for S-video to have improved quality as compared to composite video.

## Interlacing

Field interlacing is the process used to create a single video frame by overlapping two pictures where one picture provides the odd lines and the other picture provides the even lines. Field interlacing is performed to reduce bandwdith and flicker.

Figure 3.7 shows how the lines displayed on each frame are interlaced by alternating the selected lines between each image frame. In frame one, every odd line (e.g. 1,3,5, etc) is displayed. In frame two, every even line (e.g. 2,4,6, etc) is displayed. In frame 3, the odd lines are displayed. This process alternates very quickly so the viewer does not notice the interlacing operation.

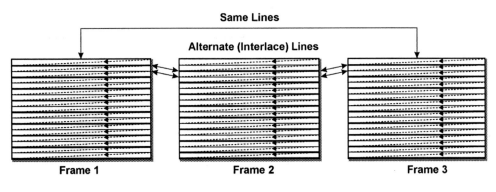

Figure 3.7, Field Interlacing

## NTSC Video

NTSC video is an established standard for TV transmission that currently in use in the United States, Canada, Japan and other countries. The abbreviation NTSC is often used to describe the analog television standard that transmits 60 fields/seconds (30 frames or pictures/second) and a picture composed of 525 horizontal scan lines, regardless of whether or not a color image is involved.

Figure 3.8 demonstrates the operation of the basic NTSC analog television system. The video source is broken into 30 frames per second and converted into multiple lines per frame. Each video line transmission begins with a burst pulse (called a sync pulse) that is followed by a signal that represents color and intensity. The time relative to the starting sync is the position on the line from left to right. Each line is sent until a frame is complete and the next frame can begin. The television receiver decodes the video signal to position and control the intensity of an electronic beam that scans the phosphorus tube ("picture tube") to recreate the display.

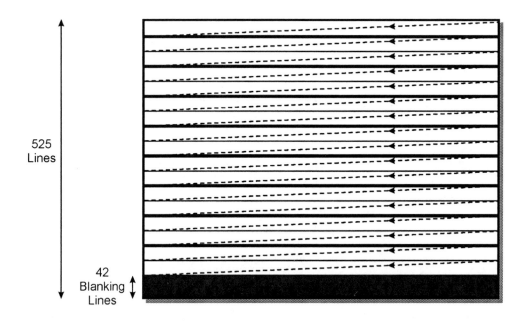

*Figure 3.8, Analog NTSC Video*

## PAL Video

Phase alternating line video is a television system that was developed in the 1980's to provide a common television standard in Europe. PAL is now used in many other parts of the world. The PAL system uses 7 or 8 MHz wide radio channels. The PAL system provides 625 lines of resolution (50 are blanking lines).

## SECAM Video

SECAM is an analog color TV system that provides 625 lines per frame and 50 fields per second. This system is similar to PAL and is used in France, the former USSR, the former Eastern Block countries, and some Middle East countries. In order to transmit color the information is transmitted sequentially on alternate lines as a FM signal.

# Digital Video

Digital video is a sequence of picture signals (frames) that are represented by binary data (bits) that describe a finite set of color and luminance levels. Sending a digital video picture involves the conversion of an image into digital information that is transferred to a digital video receiver. The digital information contains characteristics of the video signal and the position of the image (bit location) that will be displayed.

Display formatting is the positioning and timing of graphic elements on a display area (such as on a television or computer display). Because the human eye is more sensitive to light intensity than it is to color, display formats can have more intensity components than color components. Some of the common display formats include 4:2:2, 4:2:0, SIF, CIF and QCIF.

### 4:2:2 Digital Video Format

4:2:2 digital video is a CCIR digital video format specification that defines the ratio of luminance sampling frequency as related to sampling frequencies for each color channel. For every four luminance samples, there are two samples of each color channel.

Figure 3.9 shows the format of a 4:2:2 digital video on a display. In this example, a small portion of the video display has been expanded to show horizontal lines and vertical sample points of luminance (intensity) and chrominance (color). This example shows that the sample frequency for luminance for 4:2:2. is 13.5 MHz and the sample frequency for color is 6.75 MHz. This format has color samples (Cb and Cr) on every line that occur for every other luminance sample.

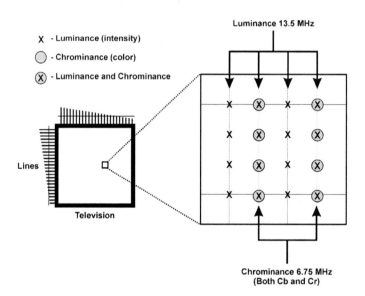

*Figure 3.9, 4:2:2 Digital Video Format*

## 4:2:0 Digital Video Format

4:2:0 digital video is a CCIR digital video format specification that defines the ratio of luminance sampling frequency as it is related to sampling frequencies for each color channel. For every four luminance samples, there is one sample of each color channel that alternates on every other horizontal scan line.

Figure 3.10 the format of a 4:2:0 digital video on a display. In this example, a small portion of the video display has been expanded to show horizontal lines and vertical sample points of luminance (intensity) and chrominance (color). This example shows that the sample frequency for luminance for

4:2:0. is 13.5 MHz and the sample frequency for color is 6.75 MHz. This format has color samples (Cb and Cr) on every other line and that the color samples occur for every other luminance sample.

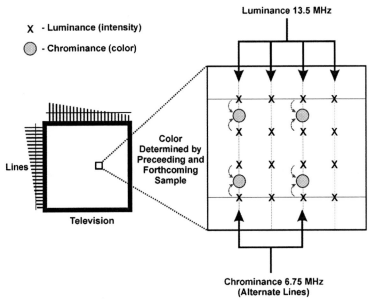

Figure 3.10, 4:2:0 Digital Video Format

## Source Intermediate Format (SIF)

Source intermediate format is a digital video format having approximately 1/2 the resolution of analog television (PAL/NTSC). SIF has a luminance resolution of 360 x 288 (625 lines) or 360 x 240 (525 lines) and a chrominance resolution of 180 x 144 (625 lines) or 180 x 120 (525 lines).

Figure 3.11 the format of a SIF digital video on a display. In this example, a portion of the video display 8 x 8 has been expanded to show horizontal lines and vertical sample points of luminance (intensity) and chrominance (color). This example shows that the sample frequency for luminance for SIF is 6.75 MHz and the sample frequency for color is 3.375 MHz. This example shows that the color samples occur on every 4$^{th}$ line and that the color samples occur for every other luminance sample.

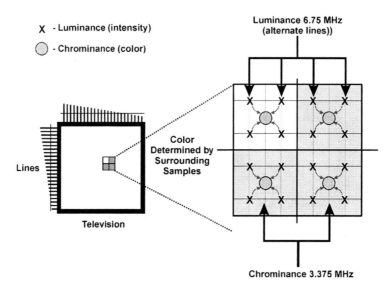

*Figure 3.11, SIF Digital Video Format*

## Common Intermediate Format (CIF)

Common interchange format (CIF) is an image resolution format that is 360 pixels across by 248 pixels high (360x248). The CIF standard is defined in the ITU H.261 and H.264 compression standards. CIF coding includes interframe prediction (using key frames and difference frames), mathematical transform coding and motion compensation.

### Quarter Common Intermediate Format (QCIF)

Quarter common interchange format (QCIF) is an image resolution format that is 180 pixels across by 144 pixels high (180x144). The QCIF standard was developed in 1990 is defined by the ITU in as H.261 and H.264 compression standards. H.261 coding includes interframe prediction (using key frames and difference frames), mathematical transform coding and motion compensation.

## Video Digitization

Video digitization is the conversion of video component signals or composite signal into digital form through the use of an analog-to-digital (pronounced A to D) converter. The A/D converter periodically senses (samples) the level of the analog signal and creates a binary number or series of digital pulses that represent the level of the optical image. Digital video may be created from another video format (e.g. analog video) or from film.

### Video Capturing

Video capturing is the process of receiving and storing video images. Video capture typically refers to capture of video images into digital form.

Digital video is usually stored in progressive format where each image in a sequence of a movie adds a new image (as opposed to a partial interlaced image) progression in a moving picture. When video is stored in progressive form, it is identified by the letter p at the end of the frame rate (e.g. 60p is 60 images per second and each frame is unique).

Figure 3.12 shows a fundamental process that can be used to digitize moving pictures or analog video into digital video. For color images, the image is filtered into red, green and blue component colors. Each of the resulting images is scanned in lines from top to bottom to convert the optical level to an equivalent electrical level. The electrical signal level is periodically sampled and converted to its digital equivalent level. This example shows that analog signals can have 256 levels (0-255) and that this can be represented by 8 bits of information (a byte).

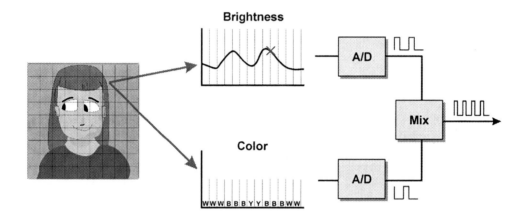

*Figure 3.12, Video Digitization*

Analog video is typically stored and transmitted in interlaced format where a single frame is converted into two interlaced fields. Analog video is converted into a progressive form by deinterlacing. Deinterlacing is the process of converting interlaced images into a form that of unique sequential images.

## Film to Video Conversion

Film to video conversion is the process of scanning film (e.g. movie) images and converting it into a sequence of video images. Because the frame rate of film is different than video, the frame rate is adapted through the use of a pulldown process.

3:2 pulldown is a process of converting film that operates at 24 frames per second to video at 60 fields per second. The 3:2 process operates by repeating a film frame image 3 times, repeating the next film frame image 2 times and repeating this process 3 repeats and 2 repeats to create 60 images (fields) per second. 3:2 pulldown is sometimes called Telecine because Telecine was a machine that performed the pulldown conversion.

Figure 3.13 shows how 24 frame per second (fps) film is converted to 60 field per second (fps) video. To create the 60 frames per second, each frame image

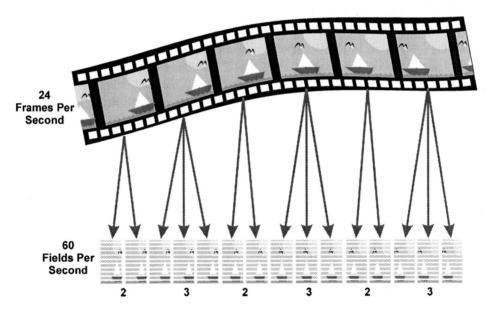

*Figure 3.13, Pulldown*

must be copied (repeated) 2.5 times. This example shows that frames with in the film are used to create 2 or 3 interlaced video fields.

# Video Compression

Video compression is the process of reducing the amount of transmission bandwidth or the data transmission rate by analog processing and/or digital coding techniques. Moving pictures can be compressed by removing redundancy within each image (spatial redundancy) or between successive images over a period of time (temporal redundancy). When compressed, a video signal can be transmitted on circuits with relatively narrow channel bandwidth or using data rates 50 to 200 times lower than their original uncompressed form.

## Spatial Compression (Image Compression)

Spatial compression is the analysis and compression of information or data within a single frame, image or section of information.

One of the common forms of spatial compression is specified by the joint picture experts group (JPEG). JPEG is a working committee under the auspices of the International Standards Organization (ISO) with the goal of defining a standard for digital compression and decompression of still images for use in computer systems. The JPEG committee has produced an image compression standard format that is able to reduce the bit per pixel ratio to approximately 0.25 bits per pixel for fair quality to 2.5 bits per pixel for high quality.

JPEG uses lossy compression methods that result in some loss of the original data. When you decompress the original image, you don't get exactly the same image that you started with despite the fact JPEG was specifically designed to discard information not easily detected by the human eye.

The JPEG committee has defined a set of compression methods that are used to provide for high-quality images at varying levels of compression up to approximately 50:1. The JPEG compression system can use compression that is fully reversible (no loss of information) or that is lossy (reversible with some loss of quality).

Lossy compression is a process of reducing an amount of information (usually in digital form) by converting it into another format that represents the initial form of information. Lossy compression does not have the ability to guarantee the exact recreation of the original signal when it is expanded back from its compressed form.

JPEG compression typically works better for photographs and reference video frames (I frames) rather than line art of cartoon graphics. This is because the compression methods tend to approximate portions of the image and the approximation of lines or sharp boundaries tends to get blurry with unwanted artifacts.

Figure 3.14 shows the basic process that can be used for JPEG image compression. This diagram shows that JPEG compression takes a portion (block) of a digital image (lines and column sample points) and analyzes the block of digital information into a new block sequence of frequency components (DCT). The sum of these DCT coefficient components can be processed and added together to reproduce the original block. Optionally, the coefficient levels can be changed a small amount (lossy compression) without significant image differences (thresholding). The new block of coefficients is converted to a sequence of data (serial format) by a zig-zag process. The data

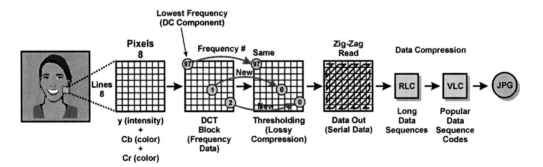

Figure 3.14, JPEG Image Compression

is then further compressed using run length coding (RLC) to reduce repetitive bit patterns and then using variable length coding (VLC) to convert and reduce highly repetitive data sequences.

## Time Compression (Temporal Compression)

Temporal compression is the analysis and compression of information or data over a sequence of frames, images or sections of information.

One of the more common forms of temporal compression used for digital is specified by the Motion Picture Experts Group (MPEG). MPEG is a working committee that defines and develops industry standards for digital video systems. These standards specify the data compression and decompression processes and how they are delivered on digital broadcast systems. MPEG is part of International Standards Organization (ISO).

Temporal video compression involves analyzing the changes that occur between successive images in a video sequence that that only the difference between the images is sent instead of all of the information in each image. To accomplish this, time compression can use motion estimation.

Motion estimation is the process of searching a fixed region of a previous frame of video to find a matching block of pixels of the same size under consideration in the current frame. The process involves an exhaustive search for many blocks surrounding the current block from the previous frame. Motion estimation is a computer-intensive process that is used to achieve high compression ratios. Block matching is the process of matching the images in a block (a portion of an image) to locations in other frames of a digital picture sequence (e.g. digital video).

Figure 3.15 shows how a digital video system can use motion estimation to identify objects and how their positions change in a series of pictures. This diagram shows that a bird in a picture is flying across the picture. In each

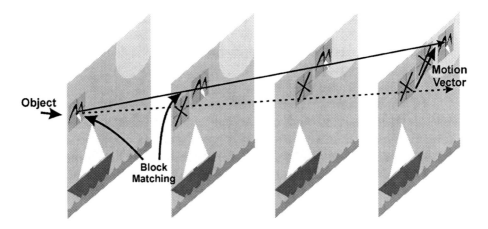

*Figure 3.15, Motion Estimation*

picture frame, the motion estimation system looks for blocks that approximate other blocks in previous pictures. Over time, the digital video motion estimation system finds matches and determines the paths (motion vectors) that these objects take.

## Coding Redundancy (Data Compression)

Coding redundancy is the repetition of information or bits of data within a sequence of data. Using data compression can reduce coding redundancy. Data compression is a technique for encoding information so that fewer data bits of information are required to represent a given amount of data. Some of the common forms of data compression used in video compression include run length encoding (RLE) and variable length encoding (VLE).

Run length encoding is a method of compressing digital information by representing repetitive data information by a notation that indicates the data that will be repeated and how many times the data will be repeated (run length).

Variable length encoding is a method of compressing digital information by representing repetitive groups of data information by code that are used to look up the data sequence along with how many times the data will be repeated (variable length).

Figure 3.16 shows how video compression may use spatial and temporal compression to reduce the amount of data to represent a video sequence. This diagram shows that a frame in a video sequence may use spatial compression by representing the graphic elements within the frame by objects or codes. The first frame of this example shows that a picture of a bird that is flying in the sky can be compressed by separating the bird image from the

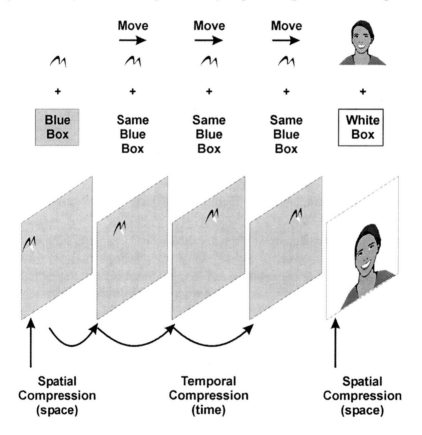

*Figure 3.16, Video Compression*

blue background and making the bird an object and representing the blue background as a box (spatial compression). The next sequence of images only needs to move the bird on the background (temporal compression).

Video images are composed of pixels. MPEG system groups pixels within each image into small blocks and these blocks are grouped into macroblocks. Macroblocks can be combined into slices and each image may contain several slices. Slices make up frames, which come in several different types. The different types of frames can be combined into a group of pictures.

## Pixels

A pixel is the smallest component in an image. Pixels can range in size and shape and are composed of color (possibly only black on white paper) and intensity. The number of pixels per unit of area is called the resolution. The more pixels per unit area provide more detail in the image.

## Blocks

Blocks are portions of an image within a frame of video usually defined by a number of horizontal and vertical pixels. For the MPEG system, each block is composed of 8 by 8 pixels and each block is processed separately.

## Macroblocks

A macroblock is a region of a picture in a digital picture sequence (motion pictures) that may be used to determine motion compensation from a reference frame to other pictures in a sequence of images. Typically, a frame is divided into 16 by 16 pixel sized macroblocks, which is also groupings of four 8 by 8-pixel blocks.

## Slice

A slice is a part of an image that is used in digital video and it is composed of a contiguous group of macroblocks. Slices can vary in size and shape.

## Frames

A frame is a single still image within the sequence of images that comprise the video. In an interlaced scanning video system, a frame comprises two fields. Each field contains half of the video scan lines that make up the picture, the first field typically containing the odd numbered scan lines and the second field typically containing the even numbered scan lines.

To compress video signals, the MPEG system categorizes video images (frames) into different formats. These formats vary from fame types that only use spatial compression (independently compressed) to frames that use both spatial compression and temporal compression (predicted frames).

MPEG system frame types include independent reference frames (I-frames), predicted frames that are based on previous reference frames (P-frames), bi-directionally predicted frames using preceding frames and frames that follow (B-Frames), and DC frames (basic block reference levels).

Intra Frames (I-Frames)

Intra frames (I-Frames) are complete images (pictures) within a sequence of images (such as in a video sequence). I-frames are used as a reference for other compressed image frames and I frames are completely independent of other frames. The only redundancy that can be removed from I frames is spatial redundancy. This means that I-frames require more data than compressed frames.

Predicted Frames (P-Frames)

Predicted frames (P-Frames) are images (pictures) within a sequence of images (such as in a video sequence) that are created using information from other images (such as from I-Frames).

Because image components are often repeated within a sequence of images (temporal redundancy), the use of P-Frames provides substantial reduction in the number of bits that are used to represent a digital video sequence (temporal data compression).

Bi-Directional Frames (B-Frames)

Bi-directional frames (B-Frames) are images (pictures) within a sequence of images (such as in a video sequence) that are created using information from preceding images and images that follow (such as from I-Frames and predicted frames P-Frames).

Because B-Frames are created using both preceding images and images that follow, B-frames offer more data compression capability than P-Frames. B-frames require the use of frames that both precede and follow the B-frames. Because B-frames must be compared to two other frames, the amount of image processing that is required for B-frames (e.g. motion estimation) is typically higher than P frames.

DC Frames (D-Frames)

A DC frame is an image in a motion video sequence that represents the DC level of the image. D frames are used in the MPEG-1 system to allow rapid viewing (e.g. fast forwarding) and are not used in other versions of MPEG.

## Groups of Pictures (GOP)

Frames can be grouped into sequences called a group of pictures (GOP). A GOP is an encoding of a sequence of frames that contain all the information that can be completely decoded within that GOP. For all frames within a

GOP that reference other frames (such as B-frames and P-frames), the frames so referenced (I-frames and P-frames) are also included within that same GOP.

The types of frames and their location within a GOP can be defined in time (temporal) sequence. The temporal distance of images is the time or number of images between specific types of images in a digital video. M is the distance between successive P-Frames and N is the distance between successive I-Frames. Typical values for MPEG GOP are M equals 3 and N equals 12.

Figure 3.17 shows how different types of frames can compose a group of pictures (GOP). A GOP can be characterized as the depth of compressed predicted frames (m) as compared to the total number of frames (n). This example shows that a GOP starts within an intra-frame (I-frame) and that intra-frames typically require the largest number of bytes to represent the image (200 kB in this example). The depth m represents the number of frames that exist between the I-frames and P-frames.

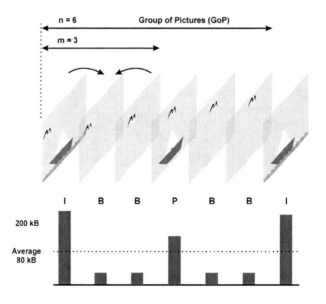

Figure 3.17, MPEG Group of Pictures (GOP)

Groups of pictures can be independent (closed) GOP or they can be relative (open) to other GOPs. An open group of pictures is a sequence of image frames that requires information from other GOPs to successfully decode all the frames within its sequence. A closed group of pictures is a sequence of image frames can successfully decode all the frames within its sequence without using information from other GOPs.

Because P and B frames are created using other frames, when errors occur on previous frames, the error may propagate through additional frames (error retention). To overcome the challenge of error propagation, I frames are sent periodically to refresh the images and remove and existing error blocks.

Figure 3.18 shows how errors that occur in an MPEG image may be retained in frames that follow. This example shows how errors in a B-Frame are transferred to frames that follow as the B-Frame images are created from preceding images.

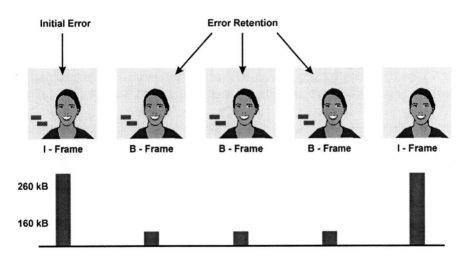

*Figure 3.18, MPEG Error Retention*

## Compression Scalability

Compression scalability is the ability of a media compression system to adapt its compression parameters for various conditions such as display size (spatial scalability), combining multiple transmission channels (layered scalability), changing the frame rate (temporal scalability) or quality of signal (signal to noise scalability).

Spatial scalability is the ability of a media file or picture image to reduce or vary the number of image components or data elements representing a picture over a given area (spatial area) without significantly changing the quality or resolution of the image.

Layered scalability is the use of multiple layers in an image that can be combined to produce higher resolution images or video. Layered compression starts with the use of a base layer may be decoded separately to provide a low resolution preview of the image or video and to reduce the decoding processing requirements (reduced complexity). An enhancement layer is a stream or source of media information that is used to improve (enhance) the resolution or appearance of underlying image (e.g. base layers).

Temporal scalability is the ability of a streaming media program or moving picture file to reduce or vary the number of images or data elements representing that media file for a particular time period (temporal segment) without significantly changing the quality or resolution of the media over time.

Signal to noise ratio scalability is the ability of a media file or picture image to reduce or vary the number of image components or data elements representing that that picture to compensate for changes in the signal to noise ratio of the transport signal.

## Advanced Video Coding (AVC/H.264)

Advanced video coding is a video codec that can be used in the MPEG-4 standard. The AVC coder provides standard definition (SD) quality at approximately 2 Mbps and high definition (HD) quality at approximately 6-8 Mbps.

The AVC coding system achieves higher compression ratios by better analysis and compression of the underlying media. The AVC system can identify and separately code objects from video sequences (object coding), it can create or represent objects in synthetic form (animated objects) and it can use variable block sizes to more efficiently represent (deblock) images with varying edges.

Object Coding

Object coding is the representation of objects (such as a graphic item in a frame of video) through the use of a code or character sequence. The types of objects that can be used in AVC system range from static background images (sprites) to synthetic video (animation).

A background sprite is a graphic object that is located behind foreground objects. Background sprites usually don't change or they change relatively slowly. Audiovisual objects are parts of media images (media elements). Media images or moving pictures may be analyzed and divided into audiovisual objects to allow for improved media compression or audiovisual objects may be combined to form new images or media programs (synthetic video).

Media elements are component parts of media images or content programs. A media element can be the smallest common denominator of an image or media program component. A media element is considered a unique specific element such as a shape, texture and size.

Animated Objects

Animated objects are graphic elements that can be created and changed over a period of time. Animated objects can be used to create synthetic video (e.g. moving picture information that is created through the use creating image components by non-photographic means).

MPEG-4 uses an efficient form (a binary form) of virtual reality modeling language (VRML) to create 3 dimensional images. This allows the MPEG system to send a mathematical model of objects along with their associated textures instead of sending detailed images that require large data transmission bandwidths.

Variable Block Sizes

Variable block sizes are groups of image bits that make up a portion of an image that vary in width and height. The use of variable block sizes allows for using smaller blocks in portions of graphic images that have lots of rapid variations (such as the edge of a sharp curve). Also, it allows larger blocks in areas that have a limited amount of variation (such as the solid blue portion of the sky).

# IP Video Transmission

IP video transmission is the transport of video (multiple images) that is in the form of data packets to a receiver through an IP data network. IP video can be downloaded completely before playing (file downloading), transferred as the video is played (streaming) or played as soon as enough of the media has been transferred (progressive downloading).

## File Downloading

File download is the transfer of a program or of data from computer server to another computer. File download commonly refers to retrieving files from a web site server to another computer.

Figure 3.19 shows how to download movies through the Internet. This diagram shows how the web server must transfer the entire media file to the media player before viewing can begin.

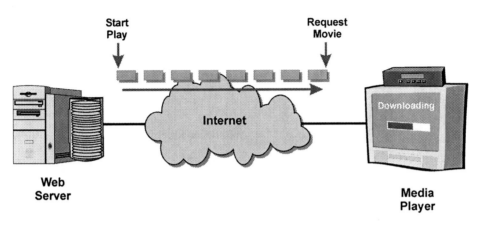

*Figure 3.19, Digital Video File Downloading*

## Video Streaming

Video streaming is the process of delivering video, usually along with synchronized accompanying audio in real time (no delays) or near real time (very short delays). Upon request, a video media server system will deliver a stream of video and audio (both can be compressed) to a client. The client will receive the data stream and (after a short buffering delay) decode the video and audio and play them in synchronization to a user.

Video streaming can be transmitted with or without flow control. When a video streaming session can obtain and use feedback from the receiver, it is called intelligent streaming. Intelligent streaming is the providing of a continuous stream of information such as audio and video content with the ability to dynamically change the characteristics of the streaming media to compensate for changes in the signal source, transmission or media application playing capabilities.

Figure 3.20 shows how to stream movies through the IP data networks. This diagram shows that streaming allows the media player to start displaying the video before the entire contents of the file have been transferred. This diagram also shows that the streaming process usually has some form of feedback that allows the viewer to control the streaming session and to provide feedback to the media server about the quality of the connection. This allows the media server to take action (such as increase or decrease compression and data transmission rate) if the connection is degraded or improved.

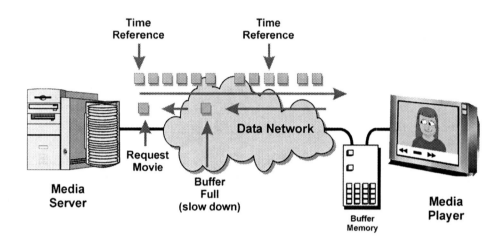

*Figure 3.20, Video Streaming*

Rate Shaping

Rate shaping is the identification, categorization and prioritization of the transfer of data or information through a system or a network to match user requirements with network capacity and service capabilities.

If the bandwidth available for a streaming media session is limited or it is decided to reduce or adjust the bandwidth for other reasons (e.g. high cost connection), rate shaping may be used to adjust the bandwidth that is assigned to a streaming connection service.

When the bandwidth available for the streaming session becomes more limited, stream thinning may be used. Stream thinning is the process of removing some of the information in a media stream (such as removing image frames) to reduce the data transmission rate. Stream thinning may be used to reduce the quality of a media stream as an alternative to disconnecting the communication session due to bandwidth limitations.

Frame Dropping

Frame dropping is the process of discarding or not using all the video frames in a sequence of frames. Frame dropping may be used to temporarily reduce the data transmission speed or to reduce the video and image processing requirements.

## Progressive Downloading

Progressive downloading is transferring of a file or data in a sequential process that allows for the using of portions of the data before the transfer is complete. An example of progressive downloading is HTTP streaming. HTTP streaming manages the sequential transferring of media files through an IP data network (such as the Internet) using HTTP commands.

## Digital Video Quality (DVQ)

Digital video quality is the ability of a display or video transfer system to recreate the key characteristics of an original digital video signal. Digital video and transmission system impairments include tiling, error blocks, jerkiness, artifacts (edge busyness) and object retention.

## Tiling

Tiling is the changing of a digital video image into square tiles that are located in positions other than their original positions on the screen.

## Error Blocks

Error blocks are groups of image bits (a block of pixels) in a digital video signal that do not represent error signals rather than the original image bits that were supposed to be in that image block.

Figure 3.21 shows an example of how error blocks are displayed on a digital video signal. This diagram shows that transmission errors result in the loss of picture blocks. In this example, the error blocks continue to display until a new image that is received does not contain the errors.

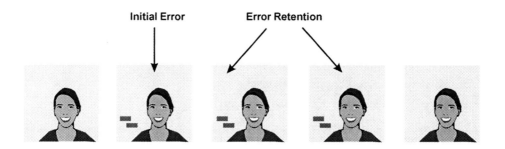

*F Figure 3.21, Digital Video Error Blocks*

## Jerkiness

Jerkiness is holding or skipping of video image frames or fields in a digital video. Jerkiness may occur when a significant number of burst errors occur during transmission that results in the inability of a receiver to display a new image frame. Instead, the digital video receiver may display the previous frame to minimize the perceived distortion (s jittery image is better than no image).

## Artifacts

Artifacts, in general, are results, effects or modifications of the natural environment produced by people. In the processing or transmission of audio or video signals, a distortion or modification produced due to the actions of people or due to a process designed by people.

Mosquito noise is a blurring effect that occurs around the edges of image shapes that have a high contrast ratio. Mosquito noise can be created through the use of lossy compression when it is applied to objects that have sharp edges (such as text).

Figure 3.22 shows an example of mosquito noise artifacts. This diagram shows that the use of lossy compression on images that have sharp edges (such as text) can generate blurry images.

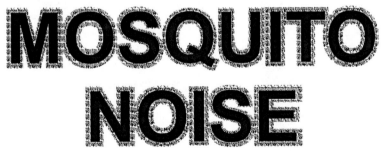

*Figure 3.22, Mosquito Noise Artifacts <*

## Object Retention

Object retention is the keeping of a portion of a frame or field on a digital video display when the image has changed. Object retention occurs when the data stream that represents the object becomes unusable the digital video receiver. The digital video receiver decides to keep displaying the existing object in successive frames until an error free frame can be received.

Figure 3.23 shows a how a compressed digital video signal may have objects retained when errors occur. This example shows that an original sequence where the images have been converted into objects. When the scene change occurs, some of the bits from image objects are received in error, which results in the objects remaining (a bird and the sail of a boat) in the next few images until an error free portion of the image is received.

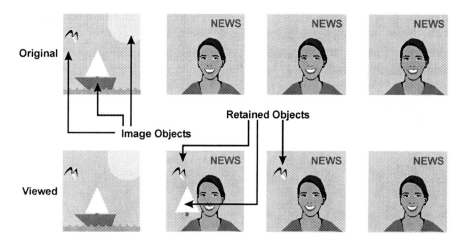

*Figure 3.23, Digital Video Object Retention*

## Streaming Control Protocols

Streaming protocols are commands, processes and procedures that are used for delivering and controlling the real-time delivery of media (such as audio and or video streaming). Streaming control protocols control the setup, playing, pausing, recording and tear down of streaming sessions. Streaming control protocols that are used for IPTV systems include real time streaming protocol (RTSP) and digital storage media command and control (DSM-CC).

### Real Time Streaming Protocol (RTSP)

Real time streaming protocol is an Internet protocol that is used for continuous (streaming) audio and video sessions. RTSP provides the control of playing, stopping, media position control (e.g. fast forward) via bi-directional (two-way) communication sessions. RTSP is defined in RFC 2326.

### Digital Storage Media Command and Control (DSM-CC)

Digital storage media command and control is an MPEG extension that allows for the control of MPEG streaming. DSM-CC provides VCR type features.

## Video Formats

Video formatting is the method that is used to contain or assign digital media within a file structure or media stream (data flow). Video formats are usually associated with specific standards like MPEG video format or software vendors like Quicktime MOV format or Windows Media WMA format.

Video formats can be a raw media file that is a collection of data (bits) that represents a flow of image information or it can be a container format that is a collection of data or media segments in one data file. A file container may hold the raw data files (e.g. digital audio and digital video) along with descriptive information (meta tags).

Some of the common video formats used in IPTV system include MPEG, Quicktime, Real media, Motion JPEG (MJPEG) and Windows Media (VC-1).

### MPEG

Motion picture experts group develops digital video encoding standards. The MPEG standards specify the data compression and decompression processes and how they are delivered on digital broadcast systems.

The MPEG system defines the components (such as a media stream or channel) of a multimedia signal (such as a digital television channel) and how these channels are combined, transmitted, received, separated, synchronized and converted (rendered) back into a multimedia format. MPEG has several compression standards including MPEG-1, MPEG-2, MPEG-4 (original) and MPEG-4 AVC/H.264.

MPEG-1 offers less than standard television resolution. MPEG-1 was designed for slow speed stored media with moderate computer processing capabilities. MPEG-2 is designed and used for television broadcaster (radio, satellite and cable TV) of standard and high definition television. The MPEG-4 specification was designed for allows for television transmission over packet data networks such as broadband Internet. The initial release of the MPEG-4 system has the same amount of video compression as MPEG-2. The MPEG-4 system was enhanced with a $2^{nd}$ type of compression called advanced video coding (AVC)/H.264 which increased the compression amount which added significant benefit to companies installing MPEG systems (more channels in less bandwidth).

## Quicktime

Quicktime is a computer video format (sound, graphics and movies) that was developed by Apple computer in the early 1990s, QuickTime files are designated by the .mov extension.

## Real Media

Real Media is a container format developed by the company "Real" and is used for streaming media files. Real Media files can use several types of coding processes. Real Media files are usually designated by the .rm file extension.

## Motion JPEG (MJPEG)

A motion JPEG is a digital video format that is only composed of independent key frames. Because MJPEG does not use temporal compression, each video frame can be independently processed without referencing other frames.

## Windows Media (VC-1)

Windows media .WM is a container file format that holds multiple types of media formats. The .WM file contains a header (beginning portion) that describes the types of media, their location and their characteristics that are contained within the media file. VC-1 is the designation for Microsoft's Windows Media Player codec by the SMPTE organization.

# Chapter 4

# Introduction to IP Audio

IP audio is the transfer of audio (sound) information in IP packet data format. Transmission of IP audio involves digitizing audio, coding, addressing, transferring, receiving, decoding and converting (rendering) IP audio data into its original audio form.

Figure 4.1 shows how audio can be sent via an IP transmission system. This diagram shows that an IP audio system digitizes and reformats the original audio, codes and/or compresses the data, adds IP address information to each packet, transfers the packets through a packet data network, recombines the packets and extracts the digitized audio, decodes the data and converts the digital audio back into its original video form.

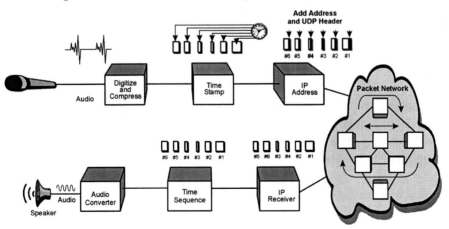

Figure 4.1, IP Audio System

## Monoral (Mono)

Monoral audio is the generation and reproduction of sound in a single channel of audio. Monoral signals are produced by a microphone that is located in a single point near the audio source that captures a sample of the audio wavefront at that location.

Monoral signals can be created from multichannel signals (such as from stereo) by combining multiple channels. However, when mono signals are created from multiple channels, a loss of fidelity (distortion) typically occurs as a result of the inability of the multiple channels to produce a signal that represents a single wavefront.

## Sterophonic (Stereo)

Stereo is the generation and reproduction of a 2-channel sound source (left and right). The use of stereo can add the ability to hear audio where the relative position of sound sources (such as instruments in a band) can be determined. Stereo signals may be transmitted through the use of independent channels or it may be sent by sending a single (mono) channel along with channel difference signal(s) that can be combined with the mono source to produce the two separate channels. Use of a single communication channel to send stereo can reduce the bandwidth requirements.

## Surround Sound

Surround sound is the reproduction of audio that surrounds the listener with sound that is provided from multiple speaker locations. The use of surround sound can allow a listener to determine the relative position of sound sources around them (such as in front and behind).

There are many variations of surround sound ranging from 3 speakers to more than 12 speakers. Generally speaking, surround sound configurations are identified by the number of full audio channels plus the number of low frequency channels. For example, surround sound 5.1 consists of 5 full audio channels and 1 low frequency channel.

Figure 4.2 shows how surround sound uses multiple speaker sources to create the effects of audio that surrounds the listener. This diagram shows a surround sound system that contains 6 channels – front left, front right, rear left, rear right, center and a low frequency enhancement (sub-audio) speaker.

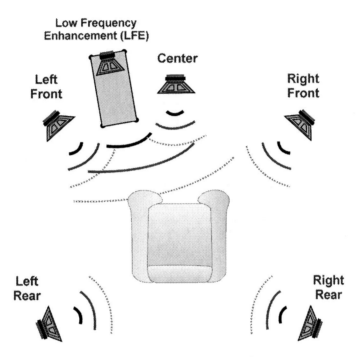

Figure 4.2, Surround Sound

# Analog Audio

Analog audio is the representation of a series of multiple sounds through the use of a signal that can continuously change (analog). This analog signal indicates the level and frequency information within the audio signal. Audio signals are transmitted through air by sound pressure waves and through wires by electrical signals. Sound pressure waves are converted to electrical signals by a microphone.

Figure 4.3 shows a sample analog signal created by sound pressure waves. In this example, as the sound pressure from a person's voice is detected by a microphone, it is converted to its equivalent electrical signal. This diagram shows that analog audio signals continuously vary in amplitude (height, loudness, or energy) as time progresses.

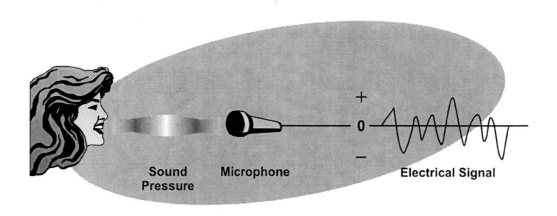

Figure 4.3, Analog Audio

Sound signals start in analog (continuous variation) form. The difference between the minimum and maximum frequency analog signals is called the frequency range and the difference between minimum and maximum signal levels that an analog signal can have its dynamic range. The dynamic range is typically expressed in decibels against a reference level. Analog signals may be processed through the use of filtering, pre-emphasis/de-emphasis, companding/expanding and other noise reduction modifications.

## Audio Filtering

Analog signal filtering is a process that changes the shape of the analog signal by restricting portions of the frequency bandwidth that the signal occupies. Because analog signals (electrical or optical) are actually constructed of many signals that have different frequencies and levels, the filtering of specific frequencies can alter the shape of the analog signal.

Filters may remove (band-reject) or allow (band-pass) portions of analog (possibly audio signals) that contain a range of high and low frequencies that are not necessary to transmit.

In some cases, additional signals (at different frequencies) may be combined with audio or other carrier signals prior to their transmission. These signals may be multiple channels (frequency multiplexing) or may be signals that are used for control purposes. If the signal that is added is used for control purposes (e.g. a supervisory tone that is used to confirm a connection exists), the control signal is usually removed from the receiver by a filter.

Figure 4.4 shows typical signal processing for an audio filter. In this example, the audio signal is processed through a filter to remove very high and very low frequency parts (audio band-pass filter). These unwanted frequency parts are possibly noise and other out of audio frequency signals that

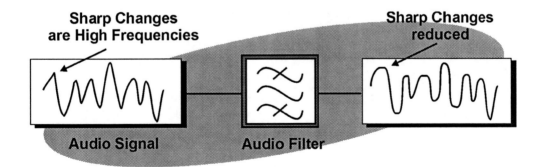

Figure 4.4, Audio Signal Filtering

could distort the desired signal. The high frequencies can be seen as rapid changes in the audio signal. After an audio signal is processed by the audio band-pass filter, the sharp edges of the audio signal (high frequency components) are removed.

## Pre-Emphasis and De-Emphasis

Pre-emphasis is the increase in the amplitude of the high-frequency components in a transmitted signal that is used to overcome noise and attenuation in the transmission system. The relative balance of high-frequency and low-frequency components is restored in a receiver by de-emphasis. De-emphasis is the alteration (decreasing) of the characteristics (e.g. amplitude) of a signal in proportion to another characteristic (such as frequency).

Figure 4.5 shows the basic signal emphasis and de-emphasis process. This diagram shows that the amount of amplifier gain is increased as the frequency of input signal is increased. This emphasizes the higher frequency

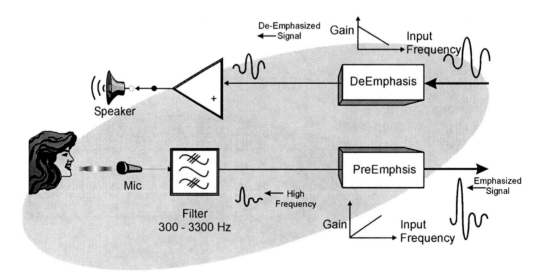

Figure 4.5, Pre-Emphasis and De-Emphasis

components as they are applied to the modulator. At the receiving end of the system, a de-emphasis system is used to reduce additional amplification to the upper frequency parts of the output signal. This recreates the shape of the original input audio signal.

## Companding and Expanding

Companding is a system that reduces the amount of amplification (gain) of an audio signal for larger input signals (e.g., louder talker). The use of companding allows the level of audio signal that enters the modulator to have a smaller overall range (higher minimum and lower maximum) regardless if some people talk softly or boldly. As a result of companding, high-level signals and low-level signals input to a modulator that may have a different conversion level (ratio of modulation compared to input signal level). This can create distortion so companding allows the modulator to convert the information signal (audio signal) with less distortion. Of course, the process of companding must be reversed at the receiving end (called expanding) to recreate the original audio signal. Expanding increases the amount of amplification (gain) of an audio signal for smaller input signals (e.g., softer talker).

Figure 4.6 shows the basic signal companding and expanding process. This diagram shows that the amount of amplifier gain is reduced as the level of input signal is increased. This keeps the input level to the modulator to a relatively small dynamic range. At the receiving end of the system, an expanding system is used to provide additional amplification to the upper end of the output signal. This recreates the shape of the original input audio signal.

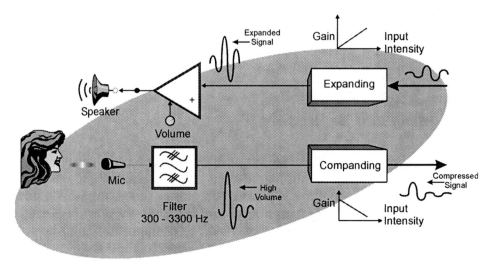

Figure 4.6, Companding and Expanding Audio

## Dolby® Noise Reduction (Dolby NR)

Dolby is an audio signal processing system that is used to reduce the noise or hiss that was invented by Ray Dolby. The original Dolby noise reduction process that was developed in 1960s used companding and expanding to adjust the dynamic range of the audio into a range that was more suitable for stored or transmitted medium. Since its original development, various enhancements to the Dolby system have been developed including Dolby A, Dolby B, Dolby C, Dolby S, Dolby SR and Digital Dolby.

Digital Dolby®

Digital Dolby (also known as Dolby 5.1) is an audio compression and channel multiplexing format that allows for the transmission of up to 6 channels (5 audio and 1 sub-audio). The sub-audio channel produces lower sound frequencies and is called the low frequency enhancement (LFE) channel.

Dolby AC-3®

Dolby AC-3 is a digital compression process that was developed by Dolby® laboratories that is commonly used in movie theaters and on DVDs.

Dolby Surround®

Dolby Surround is a version of Dolby Stereo that is used in home entertainment systems. Dolby Surround can produce similar sound effects as the Dolby Stereo system that is used in theater using a left speaker, right speaker and a rear speaker.

Dolby Pro Logic®

Dolby Pro Logic is a surround sound system that adds an additional audio channel for central speaker located in the front.

Digital Theater Sound (DTS)

Digital theater sound is a surround sound system that uses up to six digital audio channels. The DTS system includes three front sound channels (left, right and center) and a subwoofer channel. The DTS system also has separate channels for the left and right side of the theater. The DTS system uses a CD audio disk that is synchronized with the movie time codes that are stored on the film. The codes are located along each frame produce a pattern that matches codes that are stored on the CD and these codes are matched to ensure the movie and the audio are correctly synchronized.

# Digital Audio

Digital audio is the representation of audio information in digital (discrete level) formats. The use of digital audio allows for more simple storage, processing, and transmission of audio signals.

## Audio Digitization

Audio digitization is the conversion of analog audio signal into digital form. To convert analog audio signals to digital form, the analog signal is digitized by using an analog-to-digital (pronounced A to D) converter. The A/D converter periodically senses (samples) the level of the analog signal and creates a binary number or series of digital pulses that represent the level of the signal.

Audio digitization converts specific voltage levels into digital bytes of information which are based on the value (weighting) of the binary bit position. In the binary system, the value of the next sequential bit is 2 times larger.

Sampling Rate

Sampling rate is the rate at which signals in an individual channel are sampled for subsequent modulation, coding, quantization, or any combination of these functions. Sampling is the process of taking samples of an electronic signal at equal time intervals to typically convert the level into digital information. The sampling frequency is usually specified as the number of samples per unit time.

The sampling rate must be fast enough to capture the most rapid changing rates (highest frequency) contained within the analog signal. Analog signals are commonly sampled at the Nyquist rate or above. The Nyquist rate is the lowest sampling frequency that can be used for analog-to-digital conversion of a signal without resulting in significant aliasing (false signal characterizations). Normally, this frequency is twice the rate of the highest frequency contained in the signal being sampled. The typical sampling rate for conversion of analog audio ranges from 8,000 samples per second (for telephone quality) to 44,000 samples per second (for music quality).

Bit Depth

Bit depth is the number of bits that are used to represent the sample levels in an audio recording. The larger the number of bits, the more accurate the information can be represented providing for increased quality.

Bit depth is determined by the number of quantization levels. Quantization level is the number that represents the value of a sampled signal. Because a quantization number is a value that cannot represent every possible signal level, quantization of a signal results in selecting a value closest to that actual sample level. This conversion process results in quantization noise.

Quantization noise (or distortion) is the error that results from the conversion of a continuous analog signal into a finite number of digital samples that can not accurately reflect every possible analog signal level. Quantization noise is reduced by increasing the number of samples or the number of bits that represent each sample.

Figure 4.7 shows how the audio digitization process can be divided into sampling rate and quantization level. This diagram shows that an analog signal that is sampled at periodic time intervals and that the each sample gets a quantization level. This diagram shows that the quantization level may not be exactly the same value of the actual sample level. The closest quantization level is used and this causes quantization distortion.

Digital Companding

Digital audio systems may use digital companding to reduce the amount of amplification (gain) of a digital audio signal for larger input signals (e.g., louder talker). Digital companding assigns weights to bits within a byte of information that is different than the binary system. The companding process increases the dynamic range of a digital signal that represents an analog signal; smaller bits are given larger values that than their binary equivalent. This skewing of weighting values gives better dynamic range. This companding process increases the dynamic range of a binary signal by assigning different weighted values to each bit of information than is defined by the binary system.

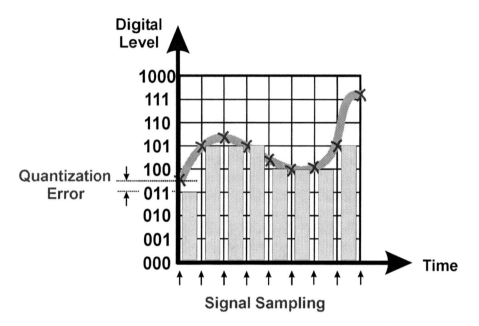

Figure 4.7, Audio Sampling and Quantization

Two common encoding laws are u-Law and A-Law encoding. u-Law encoding is primarily used in the Americas and A-Law encoding is used in the rest of the world. When different types of encoding systems are used, a converter is used to translate the different coding levels.

Dither

Dither is a random signal that is added to an analog signal during the digital audio conversion process to reduce or mask the effects of quantization noise that occurs at low levels. Without a dither signal, the digitization of an audio signal would produce a very small square wave signals. A dither signal commonly consists of white noise signal.

## Audio Format Conversion

Audio conversion is the process of changing audio signals from one format to another format. When digital signals are converted from one digital format to another, it is called transcoding. An example of transcoding is the conversion of MP3 audio coding into AAC audio coding.

# Audio Compression

Audio compression is a technique for converting or encoding audio (sound) information so that a smaller amount of information elements or reduced bandwidth is required to represent, store or transfer audio signals. Audio compression coders and decoders (codecs) analyze digital audio signals to remove signal redundancies and sounds that cannot be heard by humans.

Digital audio data is random in nature, unlike digital video, which has repetitive information that occurs on adjacent image frames. This means that audio signals do not have a high amount of redundancy, making traditional data compression and prediction processes ineffective at compressing digital audio. It is possible to highly compress digital audio by removing sounds that can be heard or perceived by listeners through the process of perceptual coding.

The characteristics and limitations of human hearing can be taken advantage of when selecting, designing and using audio signals. The human ear can hear sounds from very low frequencies (20 Hz) to approximately 20 kHz. However, the ear is most sensitive to sounds in the 1 kHz to 5 kHz.

The type of coder (type of analysis and compression) can dramatically vary and different types of coders may perform better for different types of audio sounds (e.g. speech audio as compared to music). Key types of audio coding include waveform coding, perceptual coding and voice coders.

Compression ratio is a comparison of data that has been compressed to the total amount of data before compression. For example, a file compressed to

1/4th its original size can be expressed as 4:1. In telecommunications, compression ratio also refers to the amount of bandwidth-reduction achieved. For example, 4:1 compression of a 64 kbps channel is 16 kbps.

## Waveform Coding

Waveform coding consists of an analog to digital converter and a data compression circuit that converts analog waveform signal into digital signals that represent the waveform shapes. Waveform coders are capable of compressing and decompressing voice audio, music and other complex signals such as fax or modem signals.

## Perceptual Coding

Perceptual coding is the process of converting information into a format that matches the human senses ability to perceive or capture the information. Perceptual coding can take an advantage of the inability of human senses to capture specific types of information. For example, the human ear cannot simultaneously hear loud sounds at one tone (frequency) and soft sounds at another tone (different frequency). Using perceptual coding, it would not be necessary to send signals that cannot be heard even if the original signal contained multiple audio components. Perceptual coding may remove frequency components (frequency masking) or sequences of sounds (temporal masking) that a listener cannot hear.

Frequency Masking

Frequency masking is the process of blocking, removing or ignoring specific frequency components of a signal.

Temporal Masking

Temporal masking is the process of blocking, removing or ignoring specific components of a signal that occur in a specific time period or time sequence.

Figure 4.8 shows some of the frequency and temporal masking and techniques that can be used for audio compression. This diagram shows that the compression process may remove some audio information that the listener cannot hear. This example shows that the compression process may remove small sounds that occur directly after a louder sound (temporal masking). The compressor may also remove low amplitude signals that occur simultaneously with other frequencies (frequency masking). Compression systems may also ignore small changes in sound level.

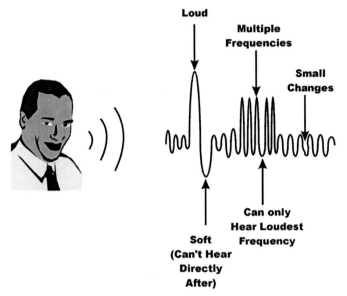

Figure 4.8, Perceptual Audio Compression

## Voice Coding

A voice coder is a digital compression device that consists of a speech analyzer that converts analog speech into its component speech parts. A speech decoder recreates the speech parts back into their original speech form. Voice coders are only capable of compressing and decompressing voice audio signals.

Figure 4.9 shows the basic digital speech compression process. In this example, the word "HELLO" is digitized. The initial digitized bits represent every specific shape of the digitized word HELLO. This digital information is analyzed and it is determined that this entire word can be represented by three sounds: "HeH" + "LeL" + "OH." Each of these sounds only requires a few digital bits instead of the many bits required to recreate the entire analog waveform.

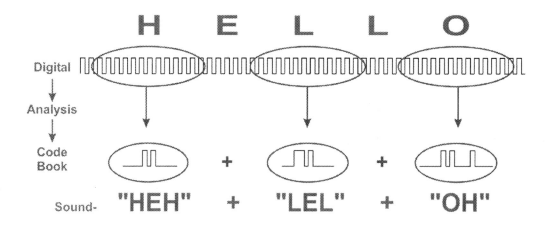

Figure 4.9, Digital Speech Compression

The type of audio coder, its analysis functions, and the data transmission rate determine the quality of the audio and how much complexity (signal processing) that is required to process the audio signal. To gain an increased amount of compression, additional signal analysis and processing is usually required. Higher compression typically results in an increase in the complexity of the coding device. Higher complexity (increased signal processing) generally increases the cost and power consumption of the coding device.

The data transmission rate for a compressed audio signal is determined by the audio sampling rate, amount of bits per sample, the compression process used and the parameters selected for the compression process. For example, two channels of audio that are sampled at 44,100 samples per second and each sample produces 16 bits of data and raw uncompressed digital audio at 1.41 Mbps.

The sampling rate and coding process is a key factor in determining the audio quality or fidelity of the audio signal. Audio fidelity is the degree to which a system or a portion of a system accurately reproduces at its output the essential characteristics of the signal impressed upon its input.

The sampling rate of an audio signal is typically performed at least 2x the highest frequency contained within the audio signal. This means if you want to convert an audio signal with a frequency range up to 10 kHz, a sampling rate of at least 20k samples per second is required.

The process of audio coding results in delays in the transmission of the digital audio signal and different types of audio coders (different processing techniques) require varying amounts of time for analysis and compression. Typically, there is a tradeoff between perceptual coding and processing delay. Increasing the amount of perceptual coding (higher compression) increases the amount of time it takes to process the signal (processing delay). Some types of audio coders are designed and configured for applications that require low transmission delays (such as real time telephone communication).

Because audio coders compress information or data into codes and these codes represent tones or other audio attributes, small errors that occur during transmission can produce dramatically different sounds. As a result, errors that occur on some of the audio data bits (e.g. high volume levels or key frequency tones) can be more sensitive to the listener than errors that occur on other data bits. In some cases, error protection bits may be added to the more significant bits of the compressed audio stream to maintain the audio quality when errors occur.

Figure 4.10 shows the basic operation of an audio codec. This diagram shows that the audio coding process begins with digitization of audio signals. The next step is to analyze the signal into key parts or segments and to represent the digital audio signal with a compressed code. The compressed code is transmitted to the receiving device that converts the code back into its original audio form.

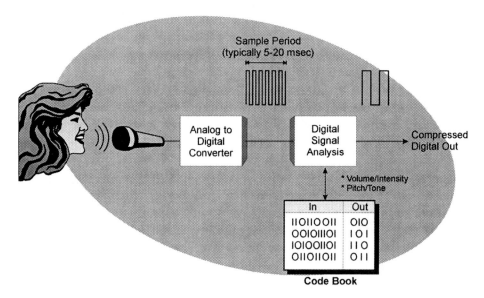

Figure 4.10, Audio Codec Operation

## IP Audio Transmission

IP audio transmission is the transport of audio (changing audio levels) that is in the form of data packets to a receiver through an IP data network. IP audio can be downloaded completely before playing (file downloading), transferring as the audio is played (audio streaming) or played as soon as enough of the media has been transferred (progressive downloading).

## Audio Streaming

Audio streaming is the continuous transfer of information that is sent through a communications network that can be represented in the audio frequency band. Upon request, a server system will deliver a stream of audio (usually compressed) to a client. The client will receive the data stream and after a short buffering delay, decode the audio and play it to a user. Internet audio streaming systems are used for delivering audio from 2 kbps (for telephone-quality speech) up to hundreds of kbps (for audiophile-quality music).

Audio streaming can be transmitted with or without flow control. The audio streaming process may use automatic rate detection or manual rate selection to control the audio streaming transmission rate. When an audio streaming session can receive feedback from the receiver, the audio server may adjust its transmission rate to compensate for changes in the data connection rates.

Figure 4.11 shows how an audio server can adjust its data transmission rate to compensate for different audio streaming data rates. This example shows how an audio server is streaming packets to an end user (for an Internet audio player). Some of the packets are lost at the receiving end of the connection because of the access device. The receiving device (a multimedia PC) sends back control packets to the audio server, indicating that the communication session is experiencing a higher than desirable packet or frame loss rate. The audio server can use this information to change the amount of audio compression and data transmission rates to compensate for the slow user access link.

## Audio Synchronization

Audio synchronization is a process that adjusts the timing of the audio signals to match the presentation of other media (such as video or a slide presentation).

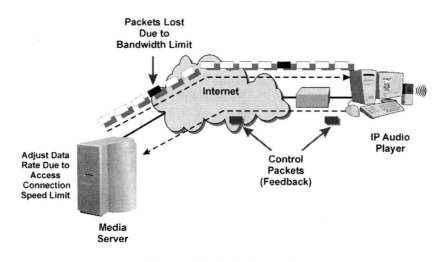

Figure 4.11, Audio Streaming

Figure 4.12 shows how audio may be synchronized with a video signal. This diagram shows that multiplexed channels can include a program clock reference (PCR) time stamps to allow all of the media streams to remain time synchronized with each other.

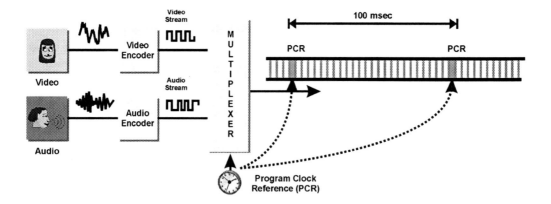

Figure 4.12, Audio Synchronization

# File Downloading

File downloading is the transfer of a program or of data from a computer server to another computer. File downloading commonly refers to retrieving files from a media server (such as web site server) to another computer.

Figure 4.13 shows how to download audio through the Internet. This diagram shows how the web audio server must transfer the entire media file to the media player before audio playing can begin.

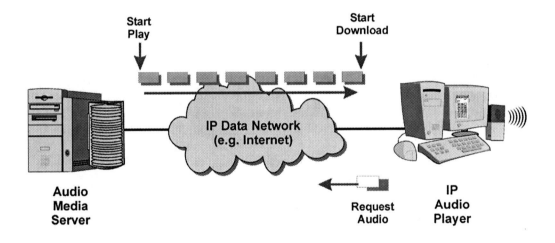

Figure 4.13, Digital Audio File Downloading

### Progressive Downloading

Progressive downloading is transferring of a file or data in a sequential process that allows for the using of portions of the data before the transfer is complete. An example of progressive downloading is HTTP streaming. HTTP streaming manages the sequential transferring of media files through an IP data network (such as the Internet) using HTTP commands that allows media to be played before the transfer is complete.

## Digital Audio Quality (DAQ)

Digital audio quality is the ability of a speaker or audio transfer system to recreate the key characteristics of an original digital audio signal. Digital audio and transmission system impairments include codec (compression) type, packet loss, packet corruption and echo (for two-way systems). Some of the measures of audio quality include fidelity, frequency response, total harmonic distortion, crosstalk, noise level and signal to noise ratio.

The type of audio coder that is used along with its compression parameters influences digital audio quality. Audio compression devices reduce the data transmission rate by approximating the audio signal and this may add distortion.

Packet loss and packet corruption errors will result in distortion or muting of the audio signal. The compression type influences the amount of distortion that occurs with packet loss or bit errors. Audio coders that have high compression ratios (high efficiency) tend to be more sensitive to packet loss and errors. Even when small amounts of errors occur in a speech coder, the result may be very different sounds (a "warble") due to the use of codebooks. Warbles are sounds that are produced during the decoding of a compressed digital audio signal that has been corrupted (has errors) during transmission. The warble sound results from the creation of different sounds than were originally sent. Muting is the process of inhibiting audio (squelching). Muting can be automatically performed when packet loss is detected.

Figure 4.14 shows some of the causes and effects of audio distortion in IP audio systems. This example shows that packet loss results in the temporary muting of the audio signal. Packet corruption results in the creation of a different altered sound than the sound that was previously transmitted. Echo results from some of the caller's audio signal being sent back (audio feedback) by the receiver of the call.

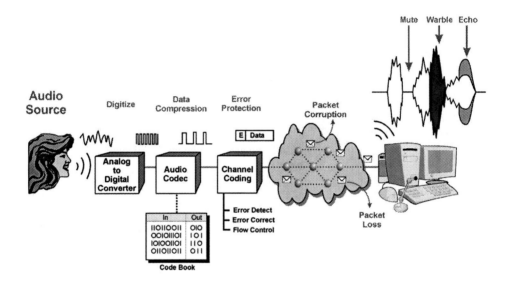

Figure 4.14, IP Audio Distortion

## Audio Fidelity

Audio fidelity is the degree to which a system or a portion of a system accurately reproduces at its output the essential characteristics of the signal impressed upon its input.

Figure 4.15 shows how to measure audio fidelity. This diagram shows that fidelity testing can identify the distortion that is added at various places in the recording, transmission and recreation of an audio signal. This diagram shows that the same reference test signal is applied to the input of the system and to a comparator. The comparator removes the original reference signal to show the amount of distortion that is added in the transmission and processing of the signal.

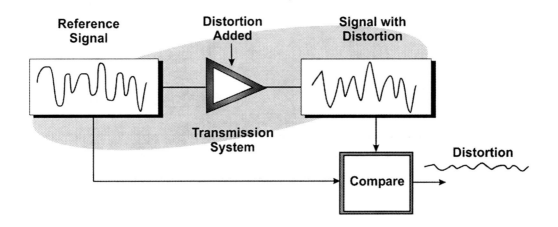

Figure 4.15, Audio Fidelity Testing

## Frequency Response (FR)

Frequency response is a measure of system linearity or performance in reproducing signals across a specified bandwidth. Frequency response is expressed as a frequency range with a specified amplitude tolerance in decibels. Frequency response in digital audio systems is typically limited to one half the sampling frequency (Nyquist limit).

## Total Harmonic Distortion (THD)

Total harmonic distortion is a ratio of the combined amplitudes of all signals related harmonically to the amplitude of a fundamental signal. THD is typically expressed as a percentage of signal level.

## Crosstalk

Cross talk is the transferring of audio from one communications channel onto another channel. Crosstalk is typically expressed as a measure of isolation in decibels between a desired channel and another channel.

## Noise Level

Noise level is a measure of the combined energy of unwanted signals. Noise level is commonly specified as a ratio (in decibels) of noise level on a given circuit as compared to decibels above reference noise level for an electrical system or decibels sound pressure level for an acoustical system.

## Signal to Noise Ratio (SNR)

Signal to noise ratio is a comparison of the information-carrying signal power to the noise power in a system. For SNR testing, a connection is setup and a test audio signal is applied to the transmitter. The energy level at the receiving end is measured and recorded. The audio test signal is then removed and the energy level at the receiving end is measured (the noise) and recorded. The difference between these two levels (commonly converted to dB) is the signal to noise ratio.

# Digital Media Formats

Digital media file formats are the sequencing and grouping of media information elements (e.g. blocks of digital audio and digital audio) within a block of data (file) or as organized on a sequence (stream) of information. File formats can range from simple linear (time progressive) sequences to indexed multiple file formatted blocks of data (containers). Some of the common multimedia file formats include media streams, indexes, container files, playlists and display control protocols.

Digital audio formats can be a raw media file that is a collection of data (bits) that represents a flow of sound information or it can be a container format that is a collection of data or media segments in one data file. A file container may hold the raw data files (e.g. digital audio and digital video) along with descriptive information (meta tags).

## Musical Instrument Digital Information (MIDI)

Musical instrument digital information is an industry standard connection format for computer control of musical instruments and devices. MIDI contains voice triggers to initiate the creation of new sounds that represent musical notes. Because MIDI files only contain trigger information and not the actual encoded media, MIDI file size is usually much smaller than other types of audio files.

## Wave Audio (WAV)

Wave audio is a waveform coding for digital audio. Wave audio files commonly have a. WAV extension to allow programs to know it is a digital audio file in Wave coding format. Wave files are uncompressed so they have a relatively large file size as compared to other types of compressed audio files such as MP3.

## Audio Interleaved (AVI)

Audio interleaved (AVI) is a Microsoft multimedia digital audio format that interleaves digital audio and digital audio frames into a common file. AVI files contain an index file of the media components.

## Advanced Streaming Index (ASX)

Advanced streaming index is a listing of the media files, their locations and the necessary parameters to setup multimedia communication sessions.

## Advanced Streaming Format (ASF)

Advanced streaming format is a Microsoft digital multimedia file format that is used to stream digital audio and digital audio. ASF is branded as Windows Media and the file extensions it may use include .WMV or .WMA. ASF files have the ability to synchronize digital audio and digital video along with managing other forms of media.

## Real Media (RM, RA, RAM)

Real Audio (RA) is Real's (the company) digital multimedia file format that is used to stream digital audio. The format was designed to be an efficient form of streaming and works with Real media players. Real audio files can have the file extension .ra.

Real Media (RM) is a digital multimedia file format that is used to stream digital audio and digital video. Real (.rm) files have the ability to synchronize digital audio along with managing other forms of media.

## MPEG

MPEG is an industry standard that allows for the use different types of audio coders. The type of coder that is selected can vary based on the application (such as playing music or speech) and the type of device the audio is being played through (such as a television or a battery operated portable media player). MPEG speech coders range from low complexity (layer 1) to high complexity (layer 3). A new version of MPEG audio coder has been created (advanced audio codec-AAC) that offers better audio quality at lower bit rates. The AAC coder also has several variations that are used in different types of applications (e.g. broadcast radio -vs.- real time telephony).

MPEG Layer 1 (MP1)

MPEG layer 1 audio is a low complexity audio compression system. MP1 was the first digital audio and it uses the precision adaptive sub-band coding (PASC) algorithm. This process divides the digital audio signal into multiple frequency bands and only transmits the audio bands that can be heard by the listener. To obtain high fidelity quality (e.g. music) with MP1 typically requires 192 kbps per audio channel.

MPEG Layer 2 (MUSICAM – MP2)

MPEG layer 2 audio is a medium complexity audio compression system which is also known as the MUSICAM system. The MUSICAM system achieves medium compression ratios, dividing the audio signal into sub bands, coding these sub bands and multiplexing them together. The MUSICAM system is used in the (DAB) digital audio broadcasting system. To obtain high fidelity quality (e.g. music) with MP2 typically requires 128 kbps per audio channel.

MPEG Layer 3 (MP3)

MPEG layer 3 is a lossy audio coding standard that uses a relatively high-complexity audio analysis system to characterize and highly compress audio signals. The MP3 system achieves high-compression ratios (10:1 or more) by removing redundant information and sounds that the human ear cannot detect or perceive. The removal of information components that cannot be detected (such as low level signals that occur simultaneously with high-level signals) is called psychoacoustic compression. To obtain high fidelity quality (e.g. music) with MP3 typically requires 64 kbps per audio channel.

The ISO/IEC Moving Picture Experts Group (MPEG) Committee standardized the MP3 codec in 1992. MP3 is intended for high-quality audio (like music) and expert listeners who have found some MP3-encoded audio to be indistinguishable from the original audio at bit rates around 192 kbps. The design of the Layer 3 (MP3) codec was constrained by backward compatibility with the Layer 1 and Layer 2 codecs of the same family.

MPEG Layer 3 Pro (MP3Pro)

MP3Pro is the Motion Picture Experts Group Layer 3 (MP3) system with spectral band replication (SBR) added to improve audio quality and/or lower the necessary data transmission rate.

## Advanced Audio Codec (AAC™)

Advanced audio codec (AAC) is a lossy audio codec standardized by the ISO/IEC Moving Picture Experts Group (MPEG) committee in 1997 as an improved but non-backward-compatible alternative to MP3. Like MP3, AAC is intended for high-quality audio (like music) and expert listeners have found some AAC-encoded audio to be indistinguishable from the original audio at bit rates around 128 kbps, compared with 192 kbps for MP3.

Advanced Audio Codec Plus (AAC Plus™)

Advanced Audio Codec Plus is a version of the AAC coder that is used to provide enhanced audio quality at high frequencies. The AACPlus system uses spectral band replication to improve the audio quality when possible. The AACPlus system has multiple streams of audio information which are composed of a base stream that can be combined with another stream to adjust (improve) the characteristics of the high-frequency audio signal components.

High Efficiency Advanced Audio Codec (HE AAC)

HE ACC is a version of the MPEG AAC system with spectral band replication (SBR) added to improve audio quality and/or lower the necessary data transmission rate.

Advanced Audio Codec Low Delay (AAC LD)

ACC LD is a version of the MPEG AAC system that is designed to provide good audio quality while providing a maximum signal processing delay that does not exceed 20 msec.

## Audio File Format (AU)

Audio file format is a media structure that contains a header and media content section. AU files can contain descriptive information, have multiple channels and the media content can be encoded in multiple encoding formats (such as PCM or ADPCM).

## Ogg

Ogg is a digital multimedia file container format that was developed by ziph.org for digital audio and digital video. The Ogg file format structure is stream oriented allowing it to be easily used for media streaming applications. The container format allows for the ability to interleave audio and video data and it includes framing structure, error detection capability along with the insertion of timestamps that can be used to synchronize streams. Ogg is an open royalty free standard which is available for anyone to use. More information about Ogg and supporting protocols can be found at www.XIPH.org.

## Windows Media Audio (WMA)

Windows media audio is an audio codec format that was created by Microsoft that works with later versions of Windows media player and integrates advanced media features such as digital rights media (DRM) control capability.

# Chapter 5

# Motion Picture Experts Group (MPEG)

Motion picture experts group (MPEG) standards are digital video encoding processes that coordinate the transmission of multiple forms of media (multimedia). MPEG is a working committee that defines and develops industry standards for digital video systems. These standards specify the data compression and decompression processes and how they are delivered on digital broadcast systems. MPEG is part of International Standards Organization (ISO).

The MPEG system defines the components (such as a media stream or channel) of a multimedia signal (such as a digital television channel) and how these channels are combined, transmitted, received, separated, synchronized and converted (rendered) back into a multimedia format.

The basic components of a MPEG system include elementary streams (the raw audio, data or video media), program streams (a group of elementary streams that make up a program) and transport streams that carry multiple programs.

Figure 5.1 shows the basic operation of an MPEG system. This diagram shows that the MPEG system allow multiple media types to be used (voice, audio and data), codes and compresses each media type, adds timing information and combines (multiplexes) the media channels into a MPEG program stream. This example shows that multiple program streams (e.g. television programs) can be combined into a transport channel. When the MPEG signal is received, the program channels are separated (demultiplexed), individual media channels are decoded and decompressed and they are converted back into their original media form.

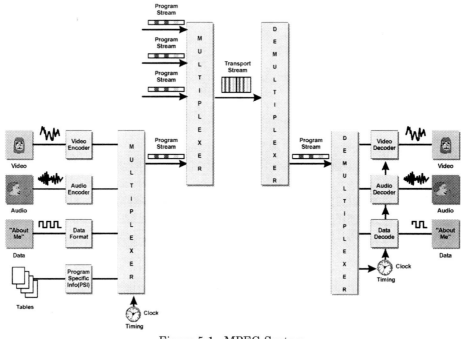

Figure 5.1., MPEG System

The MPEG system has dramatically evolved since 1991 when it was introduced primarily for use on compact disk (CD) stored media. The first version of MPEG, MPEG-1, was designed for slow speed stored media with moderate computer processing capabilities.

The next evolution of MPEG was MPEG-2, which allowed television broadcasters (such as television broadcasters, cable television and satellite television providers) to convert their analog systems into more efficient and feature rich digital television systems.

Since its introduction, the MPEG-2 system has evolved through the use of extensions to provide new capabilities. The term MPEG-2.5 is a term commonly used to describe an interim generation of MPEG technology that provides more services and features than MPEG-2 but less than the MPEG-4.

The development of an MPEG-3 specification was skipped. MPEG-3 was supposed to be created to enhance MPEG-2 to offer high definition television (HDTV). Because HDTV capability was possible using the MPEG-2 system, MPEG-3 was not released.

The next progression of MPEG technology was the release of the initial parts of the MPEG-4 specification. The MPEG-4 specification allows for television transmission over packet data networks such as broadband Internet. The initial release of the MPEG-4 system did not offer much improvement over the MPEG-2 video compression system.

To develop this more efficient video compression technology for MPEG-4, a joint video committee was created. This joint video committee was composed of members from the IETC and ITU for the purpose of analyzing, recommending and solving technical issues to create an advanced video compression specification. The result of this joint effort was the production of the advanced video coder (AVC) that provides standard definition (SD) quality at approximately 2 Mbps. This new part of MPEG-4 video compression (part 10) technology is approximately 50% more efficient (higher compression ratio) than MPEG-2 video coders. The version of AVC defined by the ITU is called H.264.

Figure 5.2 shows how the video coding developed for MPEG-4 was a joint effort between the ISO/IEC and United Nations ITU. Both groups worked together to produce the video coding standard. The ISO/IEC version is called advanced video coding (AVC) and the ITU version is called H.264.

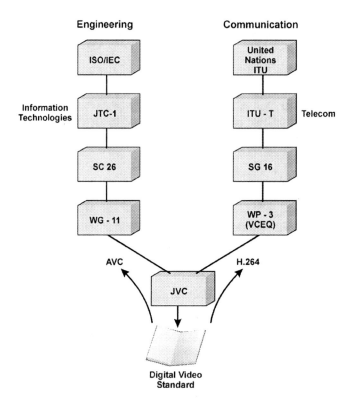

Figure 5.2., MPEG Joint Video Committee

There are other MPEG industry standards including MPEG-7, which is adds descriptions to multimedia objects and MPEG-21, which adds rights management capability to MPEG systems.

Figure 5.3 shows how MPEG systems have evolved over time. This diagram shows that the original MPEG specification (MPEG-1) developed in 1991 offered medium quality digital video and audio at up to 1.2 Mbps, primarily sent via CD ROMs. This standard evolved in 1995 to become MPEG-2, which was used for satellite and cable digital television along with DVD distribution. The MPEG specification then evolved into MPEG-4 in 1999 to permit multimedia distribution through the Internet. This example shows that work continues with MPEG-7 for object-based multimedia and MPEG-21 for digital rights management.

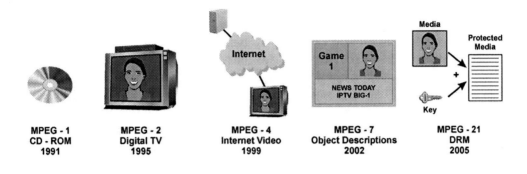

Figure 5.3., MPEG Evolution

## Digital Audio

Digital audio is the representation of audio information in digital (discrete level) formats. The use of digital audio allows for more simple storage, processing and transmission of audio signals. MPEG systems can transfer several channels of digital audio.

Because audio information is in a continuous analog form, analog audio signals are converted to digital (digitized) to allow them to be more processed and transmitted through digital networks (such as the Internet).

To convert analog audio signals to digital form, the analog signal is digitized by using an analog-to-digital (pronounced A to D) converter. The A/D converter periodically senses (samples) the level of the analog signal and creates a binary number or series of digital pulses that represent the level of the signal. The typical sampling rate for conversion of analog audio signals ranges from 8,000 samples per second (for telephone quality) to 44,000 samples per second (for music quality).

Figure 5.4 shows the basic audio digitization process. This diagram shows that a person creates sound pressure waves when they talk. These sound pressure waves are converted to electrical signals by a microphone. The bigger the sound pressure wave (the louder the audio), the larger the analog signal. To convert the analog signal to digital form, the analog signal is periodically sampled and converted to a number of pulses. The higher the level of the analog signal, the larger the numbers of pulses are created. The number of pulses can be counted and sent as digital numbers. This example also shows that when the digital information is sent, the effects of distortion can be eliminated by only looking for high or low levels. This conversion process is called regeneration or repeating. The regeneration progress allows digital signals to be sent at great distances without losing the quality of the audio sound.

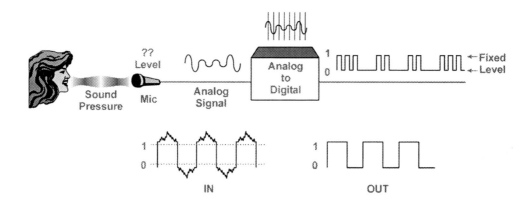

Figure 5.4., Digital Audio

When audio signals are digitized, the amount of raw digital data that is produced can be large. This presents a disadvantage and limitation when storing or transferring the raw data signals. To overcome this challenge, audio compression is used.

Audio compression is a technique for converting or encoding information so that smaller amount of information elements can be used to represent the audio signal. This allows a reduced amount of bits or lower data transmission rate to store transfer digital audio signals.

A sound system may use one (mono) or several audio signals (stereo or surround sound). Stereo is the generation and reproduction of a 2-channel sound source. Stereo signals may be transmitted through the use of independent channels or be sent by sending a single (mono) channel along with difference signal(s). Sound reprodag_MPEG_Evolution>uction that surrounds the listener with sound, as in quadraphonic recording has 4 channels. Additional audio channels include a center channel and a sub audio (very low frequency) channel. The MPEG system has the capability to send one or multiple audio channels along with other media (such as audio and video).

Figure 5.5 shows how MPEG allows multiple channels of audio. This example shows that MPEG audio may include left channel, right channel, center channel, left channel surround sound and right channel surround sound. Each of these audio channels are digitized and coded using either level 1 (low complexity), level 2 (medium complexity MUSICAM) or level 3 (high complexity MP3) coding. These channels are combined and multiplexed onto MPEG transport packets. When they are received, they are separated, decoded, and converted from digital back into their original analog form.

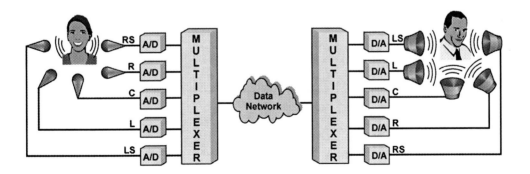

Figure 5.5., MPEG Audio

MPEG systems use audio compression to convert or encode audio (sound) information so that a smaller amount of information elements or reduced bandwidth is required to represent, store or transfer audio signals. The MPEG system supports several types of audio compression and each type can have a different amount of compression.

Figure 5.6 shows the different types of audio compression used in MPEG systems and the relative amount of compression that they can provide. This table uses a 2 channel stereo signal that is sampled at 44.1k samples per second, 16 bits per sample as a reference. The MPEG layer 1 coder can compress this signal to approximately 384 kbps (4:1 compression). The MPEG layer 2 coder can compress the signal to 192 kbps (8:1 compression). The MP3 coder can compress the signal to 128 kbps (12:1 compression). The AAC coder can compress the signal to 96 kbps (16:1 compression).

| | MPEG Layer 1 | MPEG Layer 2 | MPEG Layer 3 (MP3) | AAC |
|---|---|---|---|---|
| Raw Data Rate (stereo @ 44.1 ksamples/sec) | 1.5 Mbps | 1.5 Mbps | 1.5 Mbps | 1.5 Mbps |
| Compressed Data Rate | 384 kbps | 192 kbps | 128 kbps | 96 kbps |
| Typical Compression | 4:1 | 8:1 | 12:1 | 16:1 |

Figure 5.6., MPEG Audio Compression Comparison

# Digital Video

Digital video is a sequence of picture signals (frames) that are represented by binary data (bits) that describe a finite set of luminance (brightness) and chrominance (color) levels. Sending a digital video sequence involves the conversion of a sequence of images to digital information (e.g. through the use of image scanning) that is transferred to a digital video receiver. The digital information contains characteristics of the video signal and the position of the image that will be displayed. The formats of digital video can vary and the MPEG systems are designed to carry various forms of digital video.

Video images are composed of image elements (pixels) that contain brightness (intensity) and color (chrominance) information. Intensity or luminance is the amount of visible optical energy (intensity) and is measured in Lumens. Chrominance is the portion of the video signal that contains the color information (hue and saturation).

Video signals can be converted into digital form by either converting the combined (composite) analog video signal (intensity and chrominance levels) or by representing each image (frame) of the video by its digital pixel elements. There are several ways that digital images and video may be processed (such as progressively sending images or interlacing adjacent images). The MPEG system was designed to allow the transmission of several different digital video formats.

Figure 5.7 shows a basic process that may be used to digitize images for pictures and video. For color images, each line of image is divided (filtered) into its color components (red, green and blue components). Each position on filtered image is scanned or sampled and converted to a level. Each sampled level is converted into a digital signal.

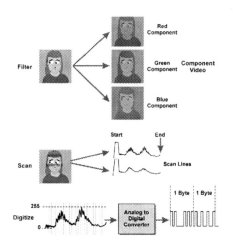

Figure 5.7., Video Digitization

Digital video compression is the reduction of the number of digital bits required to represent a video signal by digital coding techniques. When compressed, a digital video signal can be transmitted on circuits with data rates relatively 50 to 200 times lower than their original uncompressed form. The MPEG system supports many types of video compression.

Figure 5.8 shows comparison of MPEG-2 and the new MPEG-4/AVC video coding system. This diagram shows that the standard MPEG-2 video compression system requires approximately 3.8 Mbps for standard definition (SD) television and 19 Mbps for high definition (HD) television. The MPEG-4 AVC video coding system requires approximately 1.8 Mbps for SD television and 6 to 8 Mbps for HD television.

| | MPEG-2 | MPEG-4/AVC or VC-1 |
|---|---|---|
| Standard Definition (SD) | 3.8 Mbps | 1.8 Mbps |
| High Definition (HD) | 19 Mbps | 6-8 Mbps |

Figure 5.8., MPEG Video Codec Comparison

# Distribution Systems

Distribution systems are the equipment, software, and interconnecting lines that are used to transfer information to users. MPEG media may be distributed through a variety of types of distribution systems including stored media (such as CD ROMs or DVDs), land based (terrestrial) television broadcast, cable television, satellite transmission or through wired or mobile packet data networks. The characteristics of these types of distribution systems vary and this requires the MPEG system to use different options to ensure the viewer obtains the media with reliable and expected characteristics.

The original MPEG system (MPEG-1) was first developed for stored media (CD ROMs) distribution systems. This type of distribution system provides for relatively high and stable transmission rates with minimum delays and low amounts of errors.

The next version of MPEG (MPEG-2) was designed for broadcast distribution systems. Broadcast systems can provide continuous high data transmission rates with small amounts of delay and low bit error rates (called Quasi error free). Examples of broadcast systems include cable systems, satellite systems and digital terrestrial television (e.g. DVB).

MPEG then evolved to provide television signals over packet data broadcast distribution systems (MPEG-4). Packet broadcast systems often provide variable data transmission rates with varying amounts of delay. Packet broadcast channels can sometimes experience high bit error rates. Examples of packet broadcast systems include broadband television (Internet television) and mobile video.

Figure 5.9 shows some of the types of distribution systems that are used to transfer MPEG. This example shows that MPEG media may be delivered on systems that range from high bandwidth, low error rate stored media systems (e.g. CD ROM) to limited bandwidth, high error rate mobile video. MPEG-1 was designed for stable error free stored media such as CD ROMs. Stored media systems have high-transfer rates and very low error rates. MPEG-2 was developed to allow transmission over broadcast networks such as satellite systems, digital terrestrial television (DTT) and cable television systems. These broadcast systems have some errors and small delays. The MPEG-4 system was developed to allow transmission over packet data networks such as the Internet. These packet data systems usually have variable data transmission rates, unpredictable delays and may experience significant amounts of burst error rates.

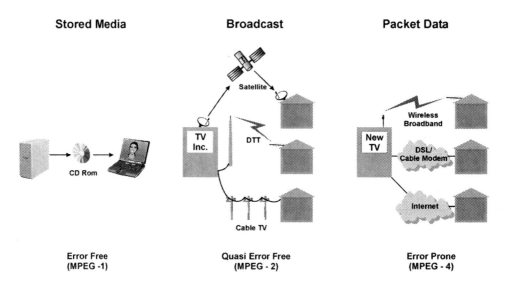

Figure 5.9., MPEG Distribution Systems

## Media Streams

A media stream is a flow of information that represents a media signal (such as digital audio or digital video). A media stream may be continuous (circuit based) or bursty (packetized). MPEG systems are composed of various types of streams ranging from the basic raw data stream (elementary streams) to stream that contain a single television video (a program stream) or a stream that combines multiple programs (transport streams).

The key elements to streaming in the MPEG system include combining multiple packet streams into a single program or transport stream, to add the time reference information into the streams and to manage the buffers required to receive and process the elementary streams.

Figure 5.10 shows the basic process of streaming video programs through a packet data network. This diagram shows that media streaming involves converting media into a stream of packets, periodically inserting time references and controlling temporary buffer sizes. .

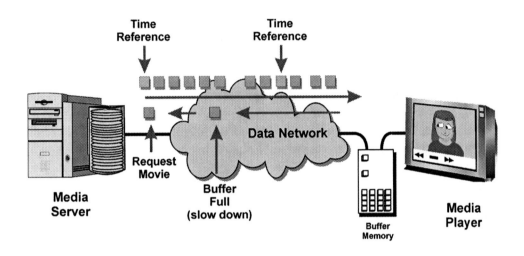

Figure 5.10., Media Streaming

## Elementary Stream (ES)

Elementary streams are the raw information component streams (such as audio and video) that are part of a program stream. MPEG system divides a multimedia source component into an elementary stream (ES). Elementary streams may be video, audio or data and there may be several elementary streams for each type of media (such as multiple audio channels for surround sound).

## Packet Elementary Stream (PES)

A packet elementary stream is a raw information component stream (such as audio and video) that has been converted to packet form (a sequence of packets). This packetization process involves dividing (segmenting) a group of bits in an elementary stream and adding packet header information to the data. This packet header includes a packet identification code (PID) that uniquely identifies the packetized elementary stream from all other packetized elementary streams that are transmitted. PES packets are variable length packets that have a length limited determined by 16 bits length field in the header of each packet.

PES streams may include time decoding and presentation time stamps that help the receiver to decode and present the media. Decoding time stamps are the insertion of reference timing information that indicates when the decoding of a packet or stream of data should occur. A presentation time stamp is reference timing values that are included in MPEG packet media streams (digital audio, video or data) that are used to control the presentation time alignment of media.

## Program Stream (PS)

A program stream is a combination of elementary streams (such as video and audio) that compose a media program (such as a television program). A program stream is called single program transport stream (SPTS). All the packets in a program stream must share the same time reference system time clock (STC).

The packet size for program streams can have different lengths. For media distribution systems that have a low error rate, longer packets may be used. For media distribution systems that have medium to high error rates (such as radio transmission or Internet systems), shorter packet lengths are typically used.

## Transport Stream (TS)

Transport Streams are the combining (multiplexing) of multiple program channels (typically digital video channels) onto a signal communication channel (such as a satellite transponder channel). A MPEG transport stream (MPEG-TS) is also called a multi-program transport stream (MPTS).

MPEG transport streams (MPEG-TS) use a fixed length packet size and a packet identifier identifies each transport packet within the transport stream. A packet identifier in an MPEG system identifies the packetized elementary streams (PES) of a program channel. A program (such as a television show) is usually composed of multiple PES channels (e.g. video and audio).

Because MPEG-TS carry multiple programs, to identify the programs carried on a MPEG-TS, a program allocation table and program mapping table is periodically transmitted which provides a list of the programs contained within the MPEG-TS. These program tables provide a list of programs and their associated PIDs for specific programs which allows the MPEG receiver/decoder to select and decode the correct packets for that specific program.

MPEG transport packets are a fixed size of 188 bytes with a 4-byte header. The payload portion of the MPEG-TS packet is 184 bytes. The beginning of a transport packet includes a synchronization byte that allows the receiver to determine the exact start time of the packet. This is followed by an error indication (EI) bit that identifies there was an error in a previous transmission process. A payload unit start indicator (PUSI) flag alerts the receiver if the packet contains the beginning (start) of a new PES. The transport priority indicator identifies if the packet has low or high priority. The 13 bit packet identifier (PID) is used to define which PES is contained in the packet. The scrambling control flag identifies if the data is encrypted. An adaptation field control defines if an adaptation field is used in the payload of the transport packet and a continuity counter maintains a count index between sequential packets.

Figure 5.11 shows an MPEG transport stream and a transport packet structure. This diagram shows that an MPEG-TS packet is fixed size of 188 bytes including a 4-byte header. The header contains various fields including an initial synchronization (time alignment) field, flow control bits, packet identifier (which PES stream is contained in the payload) and additional format and flow control bits.

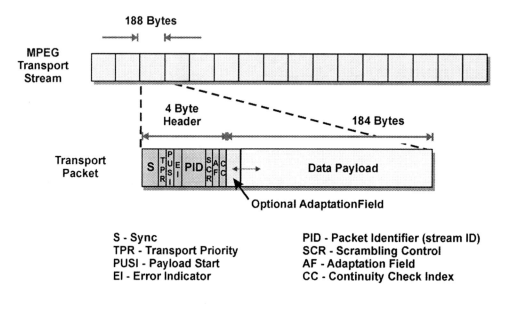

Figure 5.11., MPEG Transport Stream (MPEG-TS) Packet

PES packets tend to be much longer than transport packets. This requires that the PES packets be divided into segments so they can fit into the 184-byte payload of a transport packet. Each packet in the transport stream only contains data from a single PES. Because the division of PES packets into 184-byte segments will likely result in a remainder portion (segment) that is not exactly 184 bytes, an adaptation field is used to fill the transport packet. An adaptation field is a portion of a data packet or block of data that is used to adjust (define) the length or format of data that is located in the packet or block of data.

Figure 5.12 shows how PES packets are inserted into a MPEG transport stream. This example shows how a video and an audio packet elementary stream may be combined onto an MPEG-TS. This example shows that each of the PES packets is larger than the MPEG transport stream packets. Each PES packet is divided into segments that fit into the transport stream packets.

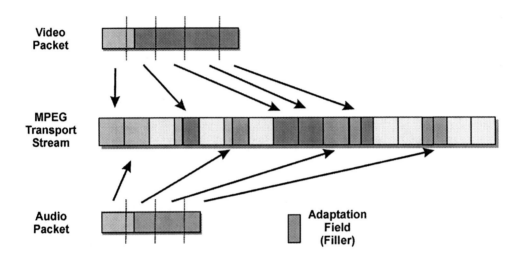

Figure 5.12., Transferring MPEG PES packets into TS Packets

## MPEG Transmission

MPEG transmission is the process of combining, sending and managing the transmission of multiple forms of information (multimedia). A program stream is a combination of elementary streams (such as video and audio) that compose a media program (such as a television program).

A multi-program transport stream is the combining (multiplexing) of multiple program channels (typically digital video channels) onto a signal communication channel (such as a satellite transponder channel). These channels are statistically combined in such a way that the bursty transmission (high video activity) of one channel is merged with the low-speed data transmission (low video activity) with other channels so more program channels can share the same limited bandwidth communication channel.

Figure 5.13 shows how MPEG transmission can be used to combine video, audio, and data onto one data communication channel. This example shows that multiple types of media signals are digitized and compressed and sent to a multiplexer. The multiplexer combines these signals and their associated time reference (clock) into a single program transport stream (SPTS). When the SPTS is received, a demultiplexer separates each of the media signal streams. Each media stream is decoded, adjusted in time sequenced using the reference clock and converted back into their original media form.

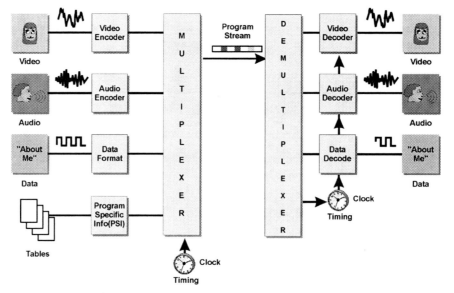

Figure 5.13.., Basic MPEG Multiplexing

## Packet Sequencing

Packet sequencing is the process of organizing packets into a sequence that is in a usable format for the system. MPEG packet sequencing from the encoder is not necessarily the same sequence that is required by the decoder. The decoding and presentation times for frames may not be the same because some of the frames may be created from future frames. B frames must be created from P frames and I frames and P frames are created from I frames.

To identify when frames should be presented, time stamps are in inserted. The MPEG system has many types of time stamps including system time stamp (STS), decoding time stamps (DTS) and presentation time stamps (PTS). MPEG system time stamps may be mandatory (such as STS) or optional time stamps (such as DTS and PTS).

Figure 5.14 shows how the encoder packet sequence differs from the decoder packet sequence. This example shows an encoder that starts providing a video sequence with a reference I frame (1). B frames (2,3) that are created from the I frame (1) and the future P frame (4) follow this. B frames (5,6) that are created follow the P frame (4) and future I frame (7). This diagram shows that because the B frames are created from I frames and P frames, the sequence of packets provided from the decoder are different. The decoder must produce the I frames and P frames before it can produce the B frames. This example shows that the output of the decoder is I frame (1) and P frame (4) which are used to produce B frames (2,3) and P frame (4) and I frame (7) are used to produce B frames (5,6).

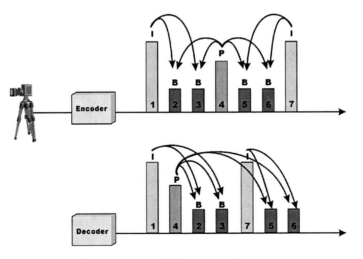

Figure 5.14.., MPEG Packet Sequencing

## Channel Multiplexing

Channel multiplexing is a process that divides a single transmission path into several parts that can transfer multiple communication (voice and/or data) channels. Multiplexing may be frequency division (dividing into frequency bands), time division (dividing into time slots), code division (dividing into coded data that randomly overlap), or statistical multiplexing (dynamically assigning portions of channels when activity exists).

A FlexMux is a set of tools that are used by a multimedia system (such as MPEG) that allows for the combining of multiple media sources (such as video and audio) so that the media streams are combined and resynchronized back into its original composite form.

Figure 5.15 shows how MPEG transmission can be used to combine video, audio, and data onto one packet data communication channel. This example shows that multiple types of signals are digitized and converted into a format suitable for the MPEG packetizers. This example shows a MPEG channel that includes video, audio, and user data for a television message. This

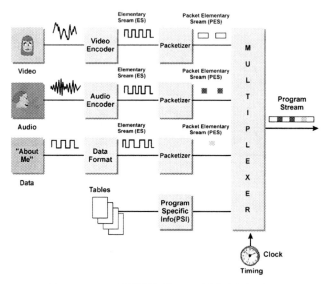

Figure 5.15., MPEG Channel Multiplexing

example shows that each media source is packetized and sent to a multiplexer that combines the channels into a single transport stream. The multiplexer also combines program specific information that describes the content and format of the media channels. The multiplexer uses a clock to time stamp the MPEG information to allow it to be separated and recreated in the correct time sequence.

## Statistical Multiplexing

Statistical multiplexing is the process of transferring communication information on an as-needed statistical basis. For statistical multiplexing systems, connections can be initiated and maintained according to anticipated activity need. Time slots or codes on a main transmission facility dynamically allocate each communication channel. This allows a communication system to operate more efficiently based by transferring information only when there is activity (such as voice or video signals).

Program channels combined on a MPEG-TS may be statistically combined in such a way that the bursty transmission (high video activity) of one channel is merged with the low-speed data transmission (low video activity) with other channels so more program channels can share the same limited bandwidth communication channel.

A statistical multiplexer can analyze traffic density and dynamically switch to a different channel pattern to speed up the transmission. At the receiving end, the different signals are merged back into individual streams.

Figure 5.16 shows how multiple MPEG channels may be combined using statistical multiplexing to help average the bandwidth usage. This example shows 3 MPEG video channels that have variable bandwidth due to high video activity periods (action scenes with high motion). The combined data rate is shown at the bottom. The combined data rate has a peak data rate that is larger than the 10 Mbps transmission channel can allow. As a result, one or more of the input MPEG channels must use a higher compression rate (temporary lower picture quality).

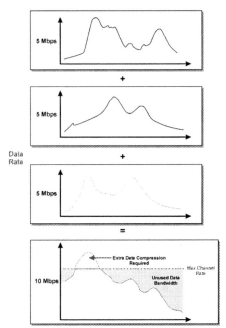

Figure 5.16., MPEG Statistical Multiplexing

# MPEG Program Tables

MPEG tables are groups of structured information that describe programs, program components or other information that is related to the delivery and decoding of programs. MPEG tables can be used by electronic programming guides (EPG) to inform the user of the available channels. The EPG is the interface (portal) that allows a customer to preview and select from possible lists of available content media. EPGs can vary from simple program selection to interactive filters that dynamically allow the user to filter through program guides by theme, time period, or other criteria.

There are many types of MPEG program tables and the more common tables contain listings of programs in a transport channel (PAT), program components (video and audio streams) and conditional access information (to enable decryption and decoding).

## Program Allocation Table (PAT)

A program allocation table contains the identification codes and system information associated with programs that are contained in a transport stream. The PAT is usually sent every 20 msec to 100 msec to allow the receiver to quickly acquire a list of available programs.

## Program Map Table (PMT)

Program map table contains information that identifies and describes the components (such as the video and audio elementary streams) that are parts of a program (such as a television show). Using the PIDs in the PMT, the receiver can select and combine the different media components to recreate the television program.

## Conditional Access Table (CAT)

A conditional access table holds information that is used by an access device (such as a set top box with a smart card) to decode programs that are part of a conditional access system (e.g. on-demand programs). If any of the programs have conditional access control, the CAT table is transmitted on PID 01. The CAT table provides the packet identifier (PID) channel code that provides the entitlement management messages (EMM) to the descrambler assembly.

## Private Tables

Private tables are user defined data elements that are sent along with broadcast programs (such as a television show).

Figure 5.17 shows how the MPEG system uses tables to describe programs and the media streams that are contained within the programs. This diagram shows that a program allocation table (PAT) is typically sent every 20 msec to 100 msec. The PAT provides the information necessary to obtain the program map table (PMT) that a viewer may want to receive. The PMT contains the information necessary to receive and decode a specific program and its media components (elementary streams). If a program has conditional access rights management associated with it, a conditional access table (CAT) will be included in the MPEG program stream. This example shows that the receiver first obtains the PAT (step 1) to obtain the PID of the PMT. After the receiver has obtained the PMT (step 2), it will use the information in the table to obtain and classify the elementary stream packets (step 3).

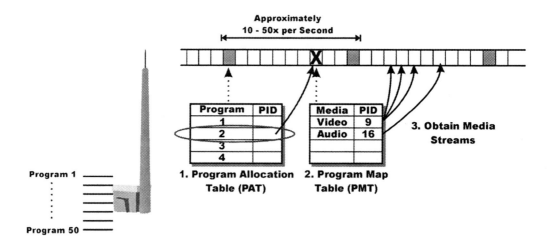

Figure 5.17, MPEG Program Tables

# Video Modes

Video modes are condition of operation that is used to transfer video images. The MPEG system can represent video images in the form of individual images (frame mode) or interlaced images (field mode).

Because some television media is stored in analog video format that use interlacing, this video must be converted into a format that can be sent on a digital broadcast system. Converting interlaced video may be performed using frame mode (alternating frames), field mode or mixed mode.

## Frame Mode

Frame mode is the process of sending video as separate images where each image (a frame) in a sequence of a movie adds a complete image. Frame mode is also known as progressive mode.

## Field Mode

Field mode is the process of sending video as images in pairs where each adjacent image (a field) of the pair contains information (image lines) that completes the image in the alternate field. Field mode for MPEG was created to allow for the sending of video that is in interlaced form.

Interlaced video is a sequence of images (video) that uses alternating graphic lines (e.g. odd and even) to represent each picture scan (fields). Analog television systems use interlacing to decrease the bandwidth required (less picture information) and increase the image presentation rate (increased images per second) to reduce flicker effects.

## Mixed Mode

Mixed mode is a process of converting interlaced video images (fields) into progressive video images (frames).

# Media Flow Control

Media flow control is the hardware and/or software mechanism or protocol that coordinates the transfer of information between devices. Flow control can be used to adjust the transmission rates between devices when one communication device cannot receive the information at the same rate as it is being sent. This can occur when the receiver requires extensive processing and the receiving buffers are running low.

For MPEG systems, media flow control primarily allows for the adjustment of transmission parameters. For one-way systems (such as MPEG-2 broadcast), flow control may be managed by the transmitting end through the use of variable compression rates or for two-way MPEG systems such as MPEG-4, information may be exchanged between the receiver and the transmitter to adjust the data transmission parameters.

## Quantizer Scaling

Quantizer scaling is the process of changing the quantizer threshold levels to adjust the data transmission rates from a media encoder. The use of quantizer scaling allows an MPEG system to provide a fixed data transmission rate by adjusting the amount of media compression.

MPEG image blocks are converted into their frequency components through the use of discrete cosine transform (DCT). The DCT converts an image map into its frequency components (from low detail to high detail). Each frequency component converted (quantized) into a specific value (coefficient). The accuracy of each of these quantized values determines how close the image block represents the original image.

Because many of the frequency components hold small values (small amounts of detail), it is possible to reduce the amount of data that represents a block of an image by eliminating the fine details through the use of thresholding. Thresholds are value that must be exceeded for an event to occur or for data to be recorded. Quantizer scaling uses an adjustable threshold level that determines if the level of frequency component should be included in the data or should a 0 level (no information) be transmitted in its place.

Figure 5.18 shows how MPEG system can use quantizer scaling to control the data rate by varying the amount of detail in an image. This example shows that an image is converted into frequency component levels and that each component has a specific level. This example shows that setting the quantizer level determines if the coefficient data will be sent or if a 0 (no data) will be used in its place.

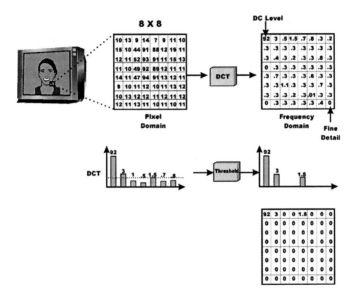

Figure 5.18., MPEG Quantizer Scaling

## Bit Rate Control

Bit rate control is a process of setting and/or adjusting the rate of a process or transmission. MPEG systems may use bit rate control to adjust the data rate of a MPEG program to match the specific capabilities or requirements of a transmission line.

When significant amounts of motion (e.g. action scenes) occur in the video, the encoded data transmission rate increases. If the data transmission rate exceeds the available transmission rate, quantizer scaling can be used to reduce the data transmission rate.

Increasing the quantizer threshold level reduces the amount of data that is sent with each block. This also reduces the detail and accuracy of each block that is transmitted. This is why digital television programs can display blockiness during periods of rapid changes in video (e.g. action scenes).

To perform bit rate control, the MPEG transmitter contains a bit rates sensing mechanism. When the bit rate from the digital video encoder starts to exceed the allowable amount (e.g. the maximum data transmission rate of the transmission channel), it increases the quantizer scaling level. The increased quantizer scaling level reduces the detail of the blocks (increased threshold level) and this reduces the data rate. When the data rate starts to decline (e.g. less changes in video), the quantizer scaling can be reduced adding detail back into the digital video.

Figure 5.19 shows how an MPEG system may use quantizer scaling to adjust the data transmission bit rate in an MPEG system. This diagram shows that the MPEG encoder is required to maintain a data transmission rate of 4 Mbps. When the data transmission rate increases, the quantizer-scaling factor is increased to reduce the data transmission rate. The trade-off of higher compression is increased blockiness in the image.

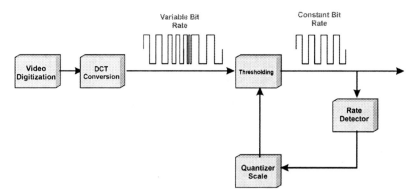

Figure 5.19., MPEG Bit Rate Control

# Buffering

Buffering is an allocation of memory storage that set aside for temporary storage of data. The use of memory buffers allows data transfer rates to vary so that differences in communication speeds between devices do not interrupt normal processing operations.

The use of buffering allows for variations in the data rate of the digital video while maintaining a constant data transmission rate. The MPEG system has the capability to setup and manage the size of buffers in receivers to allow for variation in data transmission rates.

Figure 5.20 shows how packet buffering can be used in MPEG systems to allow for some variations in digital video rates. This diagram shows that during the conversion of a digital video signal to an MPEG format, the data rate varies. Because this MPEG system has a buffer, some of the data from time periods of high video activity can be deferred into time periods of low activity. During periods of high video activity, data from the buffer is consumed at a faster rate than the transmission line can provide which results in the buffer levels decreasing. During periods of low video activity, the data from the buffer is consumed at a slower rate than the transmission line can provide allowing the buffer level to increase.

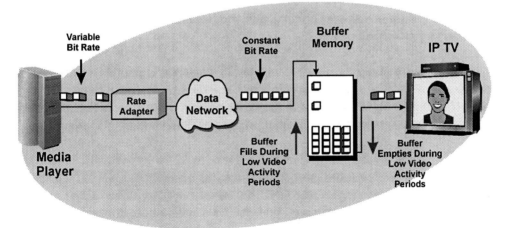

Figure 5.20., Packet Buffering

### Digital Storage Media Command and Control (DSM-CC)

Digital storage media command and control is an MPEG extension that allows for the control of MPEG streaming. DSM-CC provides VCR type features.

### Real Time Interface (RTI)

Real time interface is an extension to the MPEG system that defines an interface that allows the connection of devices to MPEG bit streams.

## Media Synchronization

Media synchronization is the process of adjusting the relative timing of media information (such as time aligning audio and video media). Media synchronization typically involves sending timing references in each media stream that can be used to align and adjust the relative timing of multiple media signals.

Media synchronization is especially important for packet based systems that can have variable amounts of delay between media sources. The variable packet transmission time may result in media components being recreated (rendered) at different times.

To provide media synchronization in MPEG systems, a program clock reference (PCR) is used. A PCR is a reference source of timing information that is used as a reference to all the media streams associated with that program. The PCR is a 42-bit field that is transmitted at least every 100 msec.

Figure 5.21 shows how MPEG can be used to time synchronize multiple media channels with each other. This diagram shows that MPEG channels include program clock reference (PCR) time stamps to allow all of the elementary streams to remain time synchronized with each other.

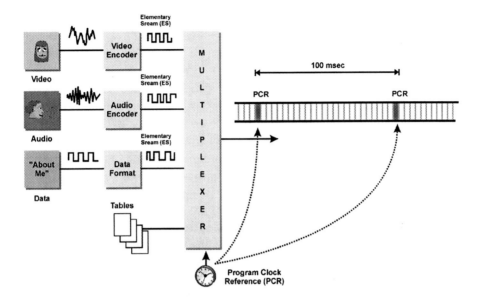

Figure 5.21., MPEG Media Synchronization

## Display Formatting

Display formatting is the positioning and timing of graphic elements on a display area (such as on a television or computer display). Display formatting may combine mixed types media such as video, animation, graphics and interactive controls on a video or television monitor.

MPEG has several protocols that can be used to position and sequence the presentation of media. These protocols include synchronized multimedia integration language (SMIL), binary format for scenes (BIFS) and active format description (AFD).

SMIL is a protocol that is used to control the user interface with multimedia sessions. SMIL is used to setup and control the operation of media files along with the placement and operation windows on the user's display. Binary format for scenes is part of the MPEG-4 standard that deals with synchronizing video and audio. AFD is a set of commands that are sent on a video bit stream that describes key areas of interest in a video or image sequence. The use of AFD allows the receiver or set top boxes to adjust or optimize the display for a viewer.

Figure 5.22 shows how MPEG can use BIFS to position and coordinate the timing of media to different areas of a display. This example shows a television that has is displaying different types of media. In the top left, a video is streaming in the window area. To the top left, an image of a new game that has been released is shown. On the bottom, text is display with interactive buttons on the right of each text line.

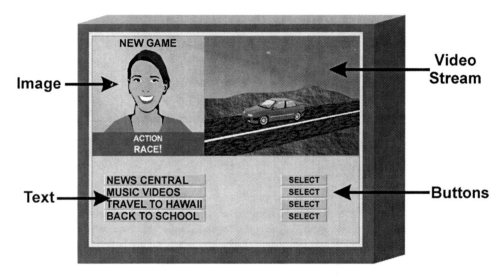

Figure 5.22., MPEG Display Formatting

# MPEG-1

MPEG-1 is a multimedia transmission system that allows the combining and synchronizing of multiple media types (e.g. digital audio and digital video). MPEG-1 was primarily developed for CD-ROM multimedia applications. Because it is primarily focused on multimedia computers, there was no defined way to process interlaced video. The compression processes used in MPEG-1 systems can compress digital video up to 52 to 1.

Part 1 of the MPEG-1 specification defines the overall system operation. It explains how the audio and video streams are combined into a single data stream and how these streams are separated, decoded and time synchronized.

Part 2 defines how digital video is coded for compressing video sequences allowing bit rates up to 1.5 Mbps. This standard defines the use of spatial (image compression) and temporal (multiple sequences) through the use of image blocks and a prediction on how these blocks will move between frames.

Part 3 defines how audio is compressed. There are 3 levels of audio compression defined in part 3 ranging from the least complex (low compression ratio) to the most complex (highest compression ratio). The most complex, level 3 is the origin of MPEG-1, part 3 (MP3).

Part 4 defines how bit streams and decoders can be tested to meet MPEG specification requirements. Part 5 provides software simulations for the MPEG-1 system.

## MPEG-2

MPEG-2 is a frame oriented multimedia transmission system that allows the combining and synchronizing of multiple media types. MPEG-2 is the current choice of video compression for digital television broadcasters as it can provide digital video quality that is similar to NTSC with a data rate of approximately 3.8 Mbps. The MPEG-2 system defines a digital video encoding process that can compress digital video up to 200 to 1.

The MPEG-2 system was developed to provide broadcast television applications so it has the ability to support progressive and interlaced video formats. The data rates for MPEG-2 systems can range from approximately 1.5 to 60 Mbps.

Part 1 of the MPEG-2 system provides overview that defines the media streams and how they are combined into programs. Part 2 describes how programs are combined onto transport streams that can be transmitted over satellite, broadcast television and cable TV networks.

Part 3 defines the how multiple channel audio can be transmitted on MPEG. The MPEG-2 audio system allows more than 2 channels of audio permitting surround sound applications. Part 4 defines how bit streams and decoders can be tested to meet MPEG specification requirements.

Part 5 provides software simulations for the MPEG-2 system. Part 6 defines the protocols that can provide media flow commands that can be used by users to start, pause and stop the play of media.

Part 7 describes the more efficient advanced audio codec (AAC) which is not backward compatible with previous MPEG audio coding systems. Part 8 was created to define how video coding could use 10 bit samples. This development of this part was discontinued.

Part 9 describes how a real interface (RTI) could be used to allow stream decoders to connect to transport streams. Part 10 contains conformance-testing processes for DSM-CC.

## MPEG-4

MPEG-4 is a digital multimedia transmission standard that was designed to allow for interactive digital television and it can have more efficient compression capability than MPEG-2 (more than 200:1 compression).

A Key feature of MPEG-4 is its ability to manage separate media components within image frames. These media objects can be independently controlled and compressed more efficiently. MPEG-4 can model media components into 2 dimensional or 3 dimensional scenes. It has the ability to sense and adjust the delivery of media dependent on the media channel type such as fairly reliable broadcast or unreliable Internet.

The overall structure and operation MPEG-4 system is described in Part 1. Part 2 defines the original video compression codec that was used in the MPEG-4 system. While this video codec offered an improvement in compression (a small amount) over the MPEG-2 video compression process, it was a relatively small improvement. Part 3 contains a set of audio codecs along with speech coding tools.

Part 4 defines the testing and conformance processes that are used to ensure devices meet the MPEG specifications. Part 5 contains reference software that can be used to demonstrate or test the operation of the MPEG-4 system.

Part 6 provides the delivery multimedia integration framework (DMIF) structure that allows a multimedia system (such as MPEG) to identify the sources of media and the transmission characters for that media source (such as from a high bandwidth low error rate DVD or through a limited bandwidth mobile telephone system). The use of DMIF allows the playback system to become independent from the sources and their transmission limitations.

Part 7 contains reference software that can be used to optimize MPEG systems. Part 8 describes how to send MPEG signals through IP networks. Part 9 provides reference hardware designs that can be used to demonstrate how to implement MPEG solutions.

Part 10 is the advanced video coding/H.264 part that provides substantial compression improvements over the MPEG-2 video compression system. Part 11 contains the binary format for scenes part of the MPEG-4 standard, which deals with synchronizing video and audio. Part 12 describes the file format that can be used for storing the media components of a program.

Part 13 contains the intellectual property management and protection is a protocol that is used in the MPEG system to enable digital rights management (DRM). Part 14 defines the container file format that can be used for MPEG-4 files. Part 15 defines the file format that can be used to store video that is compressed using the advanced video coder (AVC).

Part 16 contains the animation framework extension (AFX) set of 3D tools for interactive 3D content operating at the geometry, modeling and biomechanical level. Part 17 defines how text subtitles can be combined and timed with MPEG media. Part 18 defines how fonts can be compressed and streamed. Part 19 describes how to provide synthesized texture streams.

# MPEG-7

MPEG-7 is a system that can be used to describe the characteristics and related information about digital media objects. The MPEG-7 system is an XML based language. The MPEG-7 system uses description definition language (DDL) to describe the characteristic objects using existing (standard) and custom created definitions. Some of the standard characteristics include shape, texture and motion.

Part 1 of MPEG-7 is an overview of the key tools that are needed to describe media objects in MPEG systems. Part 2 provides the description definition language (DDL) that defines the syntax of the MPEG descriptions and how to extend (create) custom description. Part 3 explains the tools that cover visual descriptions. Part 4 explains the tools that cover audio descriptions.

Part 5 defines how to develop multimedia descriptions for objects with generic features. Part 6 contains reference software that can be used to demonstrate and test MPEG-7 systems. Part 7 provides the processes that allow testing to ensure products and software meet the requirements of the MPEG-7 system. Part 8 covers the extraction and use of descriptions. Part 9 contains the profiles and levels for MPEG-7. Part 10 describes the structure (schema) of the description definition language (DDL).

# MPEG-21

MPEG-21 is a multimedia specification that adds digital rights management capability to MPEG systems. MPEG-21 is an architecture that enables the searching, selecting, defining and managing the rights associated with digital media objects.

The MPEG-21 standard parts include digital item declaration (DID), digital item identification (DII), intellectual property management and protection (IPMP), rights expression language (REL), rights data dictionary (RDD) and digital item adaptation (DIA).

Part 2 defines digital the common set of terms and descriptions that can define a digital media object. Part 3 explains digital item identification (DII) can uniquely identify any type or portion of content. Part 4 describes the intellectual property management and protection is a protocol that is used in the MPEG system to enable digital rights management (DRM).

Part 5 defines how rights expression language (REL) protocol can be used to specify rights to content along with fees or other consideration required to secure those rights. Part is contains the rights data dictionary (RDD).

Part 7 covers digital item adaptation (DIA) which defines the semantics and syntax that may be used to adapt the format or transmission of digital items. Part 8 provides reference software that can be used to demonstrate or test the operation of the MPEG-21 system. Part 9 describes file formats.

## MPEG Profiles

MPEG profiles are a particular implementation or set of required protocols and actions that enables the providing of features and services for particular MPEG applications. These applications range from providing standard television services over a broadcast system to providing video services on a mobile wireless network. The use of profiles allows an MPEG device or service to only use or include the necessary capabilities (such as codec types) that are required to deliver media to the applications.

Profiles are created for specific applications and types of media. New profiles are constantly being requested and created. Because the types of applications of applications that use MPEG transmission can dramatically vary, MPEG has tens of profiles and there are different profile types for the MPEG-2 and MPEG-4 systems.

## MPEG-2 Profiles

MPEG-2 profiles include simple profiles (low complexity), main (standard TV), scalable profile (variable capabilities), high profile (HDTV), and 4:2:2 profile (studio quality).

### Simple Profile (SP)

The simple MPEG profile provides video for relatively simple bandwidth limited devices such as mobile telephones and personal digital assistants. The simple profile does not allow the use of bi-directional frames (B-Frames) which keeps the coding complexity low.

### Main Profile (MP)

Main profile is a common set of protocols and processes that are used to provide standard television services. The main profile used in the MPEG system allows for the use of Intra frames (I-Frames), predicted frames (P-Frames) and bi-directional frames (B-Frames). The MPEG main profile also allows for the incorporation of background sprites, interlacing and object shapes with transparency. The main profile is commonly used for broadcast television applications.

### Scalable Profile

Scalable profile is the set of MPEG protocols and processes that can provide video with varying bandwidth and performance needs. The scalable profile defines a base layer along with layers that are used to enhance the quality or performance of a video signal. The scalable profile enables the use of mobile devices that have varying capabilities of reception and display resolution.

### High Profile (HP)

High profile is the set of MPEG protocols and processes that are used for high definition television (HDTV). Although the high profile was originally developed for HDTV service, the main profile levels were expanded to enable the transmission of HDTV signals using the main profile.

## *4:2:2 Profile*

4:2:2 profile is a set of protocols and process that are used to provide high quality color video for studio production and distribution services. The 4:2:2 contains more color elements than the standard 4:2:0 YUV color format. The 4:2:2 profile allows for the use of I-frames, P-frames and B-frames. At standard resolution (720x576), its maximum data rate is 50 Mbps and at high definition (1920x1080), the maximum data rate is 300 Mbps.

Figure 5.23 shows the different types of profiles and how they are used in the MPEG-2 systems. This table shows that the key MPEG-2 profile types include simple, main, scalable, high and 4:2:2 studio profiles. Simple profiles are used for low complexity devices (such as portable media players). The main profile is used for standard broadcast television applications. Scalable profiles offer the ability to provide varying levels of service different types of devices (such as wireless devices). The high profile is used for high definition television. The 4:2:2 profile is used for studio production and distribution.

| Profile | Typical Use |
|---------|-------------|
| Simple Profile | Portable Media Players (low complexity) |
| Main Profile | Standard Television |
| Scalable Profile | Wireless Media Devices |
| High Profile | High Definition TV (HDTV) |
| 4:2:2 | Studio Production |

Figure 5.23., MPEG-2 Profile Types

## MPEG-4 Profiles

MPEG-4 profiles include audio profiles, visual profiles, MPEG-J profiles (application programming interfaces, advanced video coding (higher compression), scene distribution (dimensional control) and graphics profiles (image processing).

### Audio Profiles

Audio profiles are protocols and processes that are used to provide audio transfer and rendering applications and services. The audio profiles focus on specific applications that require tradeoffs such as the tradeoff of having a high audio compression ratio which results in increased audio processing (coding time) delays and higher sensitivity to transmission errors. Additional tradeoffs include speech coding (optimized for human voice) that can offer a high compression ratio as opposed to natural audio coding with relatively low compression ratios that can reproduce any sound.

The MPEG-4 system has several audio profiles including speech profile, synthesis profile, scalable profile, main profile, high quality audio profile, low delay audio profile, natural audio profile and mobile audio internetworking profile.

The speech profile can be used to produce a range of speech audio functions ranging from high bit rate waveform coding to very low bit rate text to speech audio coding. Text to speech conversion transforms text (ASCII) information into synthetic speech output. This technology is used in voice processing applications that require the production of broad, unrelated and unpredictable vocabularies, e.g., products in a catalog, names and addresses, etc.

Synthesis profile defines how the MPEG system can use wave table synthesis and text to speech to create audio. The scalable profile can use different bit rates and audio bandwidths to effectively provide audio that contains music and speech. Main profile contains tools for natural and synthetic audio.

High quality audio profile allows for the use of advanced audio coding (AAC) which is more efficient (has a higher data compression ratio) than the MP3 coder. The low delay audio profile uses speech coders and text to speech interfaces to provide low delay real time communication. It is a version of the advanced audio coder that reduces the amount of processing time for coding which can be used for real time two-way communication applications.

Natural audio profile contains the capabilities for natural (non-synthetic) audio coding. Mobile audio internetworking profile uses a scalable AAC with low delay to enable high quality audio coding processes for time sensitive applications (such as video conferencing).

## *Visual Profiles*

Visual profiles are protocols and processes that are used to provide video transfer and rendering services. The MPEG-4 system video profiles have capabilities that include the transfer of arbitrarily shaped objects (as opposed to rectangular shaped objects), scalable (variable quality) images, layered (images that can be enhanced) and the transfer and remote (local creation) of animated images.

Simple visual profile is a low complexity video coding system that can be used for portable video devices such as multimedia cellular telephones. Because it is a simple profile, it does not include the ability to provide interlaced video.

Simple scalable visual profile can carry over multiple sequences (temporal scalability) or over multiple areas (spatial scalability) to allow operation at different bit rates.

The fine granularity scalability profile has the ability to layer multiple levels of image quality. It starts with a base layer and ads enhancement layers to improve the resolution of the images. It can be used in systems to offer different quality levels when bandwidth is selectable or adjustable.

Advanced simple profile (ASP) can use B-frames and can use ¼ pel compensation to increase the compression ratio (it is approximately 1/3$^{rd}$ more efficient than simple visual profile). Core visual profile has the ability to define objects that have different (arbitrary) shapes. It is useful for applications that have interactivity between media objects. The core scalable profile includes improved scalability options for arbitrarily shaped objects.

Main visual profile includes the ability to process interlaced video objects. The main visual profile can be used for broadcast and interactive DVD applications.

N-bit visual profile defines how to code video objects using pixel depths that range from 4 bits to 12 bits. Advanced core profile includes the ability to process arbitrarily shaped video and still images.

Advance coding efficiency profile (ACE) adds tools that can increase the compression ratio for both rectangular and arbitrarily shaped video objects.

Simple studio profile is used to maintain the quality of video for studio distribution and editing functions. The studio simple only uses I frames. The core studio profile ads P-frames to the simple studio profile allowing a higher compression ratio at the expense of added complexity.

The simple facial animation visual profile uses a mathematical model to animate facial images. This allows for the sending of a very small amount of data (e.g. lips move a small amount) to create 3 dimensional images. The simple face and body animation profile uses a mathematical model to animate both facial and body images.

Advanced real time simple (ARTS) uses a low delay coding process along with robust error coding to provide real time two-way communication that cannot tolerate delays and which are subject to high error rates.

The scalable texture visual profile allows for the mapping of textures onto images. Basic animated 2D texture visual profile has the capability to insert textures onto 2D images. Advanced scalable texture profile can perform scalable shape coding. This includes wavelet tiling.

The hybrid visual profile includes the capability to decode synthetic and natural objects.

## Advanced Video Coding Profiles

Advanced video coding profiles are protocols and processes that are used to provide advanced video transfer and rendering capabilities to underlying MPEG services. The AVC profiles use the H.264/AVC coder that provides a compression ratio that is approximately double the MPEG-2 video coder. The AVC profiles range from low complexity baseline profile (BP) to an ultra high professional quality 4:4:4 profile.

The baseline profile (BP) is used for low complexity device such as portable media players. Extended profile (XP) is designed for streaming video over networks that may have high packet loss (such as the Internet).

The main profile was designed for broadcast (e.g. Cable TV) and storage (e.g. DVD) applications. High profile (HiP) can provide high definition television for broadcast and stored media distribution systems. High 10 profile (Hi10P) provides for increased quality allowing up to 10 bits per sample of decoded picture elements.

High 4:2:2 profile (Hi422P) is used for professional applications that require higher levels of chroma (color) elements. High 4:4:4 profile (Hi422P) adds additional chroma (color) elements and allows up to 12 bits per image element.

## MPEG-J Profiles

MPEG-J profiles are protocols and processes that are used to provide application programming interface (API) capabilities to underlying MPEG services. The MPEG-4 system has several MPEG-J profiles including personal profile and main profile.

Personal MPEG-J profile is used for low complexity portable devices such as gaming devices, mobile telephones and portable media players. Main

MPEG-J profile includes all the profiles from personal MPEG-J profiles plus APIs that can be used to select and configuring decoders along with interfaces to access service information.

## Scene Graph Profiles

Scene graph profiles are the sets of protocols and processes that are used to define a composite (mixture) of media objects and how they relate to each other in an area. Scene graph profiles can be used to define the hierarchical relationship between videos, graphic images and audio components in a 2 dimensional or 3 dimensional enviornment.

Some of the scene graphs profiles are based on the virtual reality modeling language (VRML) protocol. VRML is a text based language that is used to allow the creation of three-dimensional viewpoints, primarily for use with Web browsing.

The MPEG-4 system scene graph profiles include the basic 2D profile, simple 2D scene graph profile, complete 2D scene graph profile, core 2D profile, advanced 2D profile, main 2D profile, X3D profile, complete scene graph profile, audio scene graph proifle and 3D audio scene graph profile.

The basic 2D profile is used to define simple scenes on a 2 dimensional area. It is used for audio only or video only applications. The simple 2D scene graph profile can place media objects into a scene on a 2 dimensional area. It is a low complexity profile and it and does not define interaction with the media objects. Complete 2D scene graph profile allows for alpha blending and interactivity with the media objects. Alpha blending is the combining of a translucent foreground color with a background color to produce a new blended color.

The core 2D profile can use both audio and visual media objects. It allows for local animation and interaction. Local animation is a process that changes parameters or features of an image or object over time that is processed at the location where the animation is playing (e.g. within a television or multimedia computer).

The advanced 2D profile contains the capabilities of the basic 2D and core 2D profiles along with adding scripting capability and local interaction.

The main 2D profile ads FlexeTime model which allows input sensors and additional tools for interactive applications. It is designed to interoperate with synchronized multimedia integration language (SMIL). SMIL is a protocol that is used to control the user interface with multimedia sessions. SMIL is used to setup and control the operation of media files along with the placement and operation windows on the users display.

The X3D profile is a small footprint (limited memory and processing requirements) 3 dimensional media processing profile. It is designed to interoperate with X3D specification created by the Web3D consortium. Extensible 3D (X3D) is a storage, retreival and rendering (playback) industry standard for real time graphics media objects which can be adjusted in relative position and possibly interact with each other.

Complete scene graph profile is a combined set of 2D and 3D scene graph profiles from the binary format for scenes (BIFS) toolbox. Complete scene graph profile can be used for vitual gaming that have 3 dimensional worlds.

Audio scene graph profile is used for applications that only require audio media. The 3D audio scene graph profile describes how to position sound in a 3 dimensional environment. It allows for interaction of sounds with objects within the scene.

## *Graphics Profiles*

Graphics profiles are sets of protocols and processes that are used to define and control graphics elements that are used in scenes. Some of the graphics profiles used in MPEG-4 systems include simple 2D graphics profile, simple 2D + text profile, core 2D profile, advanced 2D profile, complete 2D graphics profile, comlete graphics profile, 3D audio graphics profile, and X3D core profile.

The simple 2D graphics profile is used for placing media objects in a scene that has only 2 dimensions. The simple 2D + text profile adds the ability to include text on the screen and to allow the text to move and be transformed (e.g. become transparent) with other media objects.

Core 2D profile is used for picture in picture (PIP), video warping, animated advertisements and logo insertion. The advanced 2D profile adds graphic user interface (GUI) capabilities along with more complex graphics control for animation.

Complete 2D graphics profile is a full set of 2D graphics control capabilities including the use of multiple shaped graphic objects. The complete graphics profile allows for the use of elevation grids, extrusions and lighitng effects to create virtual worlds.

3D audio graphics profile is used to define the acoustical properties of a scene. It includes features such as acoustics absorption, acoustic diffusion, acoustic transparency and tele-presence.

The X3D core profile includes 3 dimenstional media object processing capabilities for advanced gaming and virtual environments. It is compatible with X3D specification.

Figure 5.24 shows the different types of profiles used in the MPEG-4 systems. This table shows that the key MPEG-4 profile types include audio,

| Profile Type | Profile Uses |
|---|---|
| Audio | Audio Media Players (from simple audio only to digital cinema) |
| Visual | Various Levels of Video Support (from variable resolution to studio quality) |
| Advanced Video Coding | Various Levels of Video Support using Advanced (high compression) Video Coder |
| MPEG-J (Java) | Application Programming Interfaces |
| Scenes | Relationships between Media Objects (2D and 3D Scenes) |
| Graphics | Definition of the Graphics Elements used in Scenes |

Figure 5.24., MPEG-4 Profile Types

visual, advanced video, MPEG-J, scenes and graphics profiles. The audio profiles range from very low bit rate to high quality audio that can be used in cinemas. The visual profiles range from low resolution profiles that can be used in portable devices to high quality studio profiles. Advanced video coding profiles take advantage of the high compression AVC coder. MPEG-J profiles allow the device to have direct programming interfaces to the MPEG media and devices. Scenes profiles define the relationships between media objects. Graphics profiles define media objects and how they are positioned on displays and how they can be changed in scenes.

## MPEG Levels

MPEG levels are the amount of capability that a MPEG profile can offer. The use of levels allows products to define their quantitative capabilities such as memory size, resolution and maximum bit rates. For example, MPEG levels can range from low detail (low resolution) to very high capability (high resolution).

Low level MPEG signals have a resolution format of up to 360 x 288 (SIF). Main level MPEG signals provide for resolution of up to 720x576 (standard definition television). High level can support resolution of up to 1920x1152 (high definition). One of the more common profile combinations is the main profile at the main level (MP@ML). This combination provides television signals for standard definition (SD).

Figure 5.25 shows the different levels used in the MPEG-2 system. This table shows that low level has low resolution capability with a maximum bit rate of 4 Mbps. The main level has standard definition (SD) capability with

| Name | Lines | Pixels per Line | Bitrate (Mbps) |
|---|---|---|---|
| Low Level (LL) | 352 | 288 | 4 |
| Main Level (ML) | 576 | 720 | 15 |
| High-1440 (H-14) | 1152 | 1440 | 60 |
| High Level (HL) | 1152 | 1920 | 80 |

Figure 5.25., MPEG-2 Levels

a maximum bit rate of 15 Mbps. The high 1440 profile is a high definition format that has a maximum bit rate of 60 Mbps. The high level profile has a high definition format with 1152 lines that have 1920 pixels per line which can have a bit rate of up to 80 Mbps.

## Conformance Points

Conformance points are a combination of profiles and levels in a system (such as an MPEG system) where different products can interoperate (by conforming to that level and profile). An example of conformance points is if ability of a mobile video server to support creation and playback of a simple visual profile at level 0, any mobile phone that has these conformance points should be able to play a video with these profiles and levels.

# Chapter 6

## IP Television Systems

There are three basic types of systems that are used to provide IP television services; managed IP television; Internet television service provider (ITVSP) and private IP television systems.

## Managed IP Television Systems

Managed IP television systems provide services to customers using controlled access Internet protocol (IP) connections. Managed IP television systems allow customers to have the quality and features that are typically associated with a traditional television system (such as cable television – CATV). Because the IP television provider manages (controls) the bandwidth of the access network, it is possible to guarantee the video quality and reliability of the IP television service to the customer.

Managed IP television systems include IP television over cable, telephone ("Telco TV"), IP television over cable modem, wireless broadband or digital television over powerline where the IP television provider controls both the access network and the services.

## Internet Television Service Providers (ITVSPs)

Internet television service providers (ITVSPs) supply IP television services to their customers that are connected to broadband Internet connections ("over the top of the broadband Internet"). ITVSPs provide access screens or channel links to allow the customer to connect their viewers to media sources for a fee. ITVSPs can vary from companies that simply provide IP television links to viewers to media content providers.

Customers use IP television access devices (media players) or IP set top boxes (self contained media players) to communicate their requests for services and features to the ITVSP. The ITVSP receives their requests and determines what features they are authorized to access that particular feature or service. If the ITVSP decides to provide service, it will determine which gateway it will use to connect the viewer to the media source. The gateway will record the time and usage information and send it to the ITVSP to account for the usage (to get paid).

ITVSP companies are primarily made of computers that are connected to the Internet and software to operate call processing and other services. ITVSPs use computer servers to keep track of which customers are active (registration) and what features and services are authorized. When television channel requests are processed, the ITVSP sends messages to gateways via the Internet allowing television channels to be connected to the viewer's media player.

Because there are many television gateways located throughout the world (some television programs may have multiple gateways), gateways are commonly connected to clearinghouse companies to settle the usage charges. These clearinghouses gather all the billing details (which may be hundreds of thousands) of billing records per month for each ITVSP that has used the gateway and create invoices for each ITVSP periodically.

Because ITVSPs provide IP television signals through public broadband networks, they cannot directly control the quality of service for data transmission. While the ITVSP cannot directly control or guarantee IP television quality, if the data links in the connection from the media source to the viewer have data rates that are much higher than the viewer's data rate, the channel quality is likely to be good for the selected digital video signal.

# Private IP Television Systems

Private IP television systems are used to provide television service within a building or group of buildings in a small geographic area. Private IP television systems typically contain media servers, media gateways, and advanced television channel selection processing features such as channel selection, media control (play, stop, fast forward). Applications for private IP television systems include hotel television, community television, surveillance security and video conferencing. Private IP television systems may be directly connected to media sources (e.g. DVD video players) and/or they may be connected to other media sources through gateways.

There are several ways that private television systems (such as CATV systems) can be upgraded for IP television capability. The options range from adding IP television gateways as a media source to allow the existing television systems to receive new television channels to replacing the media sources and set top boxes with IP video devices.

## IP Television (IPTV) Networks

IPTV networks are television distribution systems that use IP networks to deliver multiple video and audio channels. IP Television networks are primarily constructed of computer servers, gateways, access connections and end user display devices. Severs control the overall system access and processing of calls and gateways convert the IP television network data to signals that can be used by television media viewers.

For many years, video (television) broadcasters had monopolized the distribution of some forms of information to the general public. This had resulted in strict regulations on the ownership, operation, and types of services broadcast companies could offer. Due to the recent competition of wide area information distribution, governments throughout the world have eased their regulation of the broadcast industry. In 1996, the United States released the Telecommunications Act of 1996 that dramatically deregulated the telecommunications industry. This allowed almost all types of communication systems to provide many new services with their existing networks including television services.

IPTV system operators link content providers to content consumers. To do this, IPTV systems gather content via a content network and convert the content to a format that it can use via a headend system. It then manages (e.g. playout) the content via an asset management system, transfers the content via a distribution network, and then the media may be displayed on a variety of viewing devices.

Figure 6.1 shows a sample IPTV system. This diagram shows the IPTV system gathers content from a variety of sources including network feeds, stored media, communication links and live studio sources. The headend converts the media sources into a form that can be managed and distributed. The asset management system stores, moves and sends out (playout) the media at scheduled times. The distribution system simultaneously transfers multiple channels to users who are connected to the IPTV system. Users view IPTV programming on analog televisions that are converted by adapter box (IP set top box), on multimedia computers or on IP televisions (data only televisions).

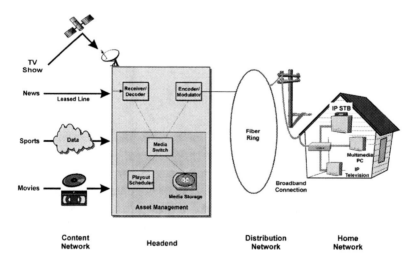

Figure 6.1, IPTV System

# Contribution Network

A contribution network is a system that interconnects contribution sources (media programs) to a content user (e.g. a television system). IPTV systems receive content from multiple sources through connections that range from dedicated high-speed fiber optic connections to the delivery of stored media. Content sources include program networks, content aggregators and a variety of other government, education and public sources.

## Connection Types

IPTV content distribution network connection types include satellite connections, leased lines, virtual networks, microwave, mobile data and public data networks (e.g. Internet).

Satellite communication is the use of orbiting satellites to relay communications signals from one station to many others. A satellite communication link includes a communication link that passes through several types of systems. These connections include the transmission electronics and antenna, uplink path, satellite reception and transmission equipment (transponder), downlink path, and reception electronics and antenna. Because satellite systems provide signal coverage to a wide geographic area, the high cost of satellites can be shared by many broadcasting companies.

Satellite content distributors that provide television programming to IPTV networks via satellite lease some or all of the transponder capacity of the satellite. Satellite content providers combine multiple programs (channels) for distribution to broadcasters.

Leased lines are telecommunication lines or links that have part or all of their transmission capacity dedicated (reserved) for the exclusive use of a single customer or company. Leased lines often come with a guaranteed level of performance for connections between two points. Leased lines may be used to guarantee the transfer of media at specific times.

Virtual private networks are private communication path(s) that transfer data or information through one or more data network that is dedicated between two or more points. VPN connections allow data to safely and privately pass over public networks (such as the Internet). The data traveling between two points is usually encrypted for privacy. Virtual private networks allow the cost of a public communication system to be shared by multiple companies.

A microwave link uses microwave frequencies (above 1 GHz) for line of sight radio communications (20 to 30 miles) between two directional antennas. Each microwave link transceiver usually offers a standard connection to communication networks such as a T1/E1 or DS3 connection line. This use of microwave links avoids the need to install cables between communication equipment. Microwave links may be licensed (filed and protected by government agencies) or may be unlicensed (through the use of low power within unlicensed regulatory limits). Microwave links are commonly used by IPTV systems to connection remote devices or locations such as a mobile news truck or a helicopter feed.

Mobile data is the transmission of digital information through a wireless network where the communication equipment can move or be located over a relatively wide geographic area. The term mobile data is typically applied to the combination of radio transmission devices and computing devices (e.g. computers electronic assemblies) that can transmit data through a mobile communication system (such as a wireless data system or cellular network). In general, the additional of mobility for data communication results in an increased cost for data transmission.

The Internet is a public data network that interconnects private and government computers together. The Internet transfers data from point to point by packets that use Internet protocol (IP). Each transmitted packet in the Internet finds its way through the network switching through nodes (computers). Each node in the Internet forwards received packets to another location (another node) that is closer to its destination. Each node contains routing tables that provide packet forwarding information. The Internet can be effectively used to privately transfer programs through the use of encryption.

In additional to gathering content through communication links, content may be gathered through the use of stored media. Examples of stored media include magnetic tapes (VHS or Beta) and optical disks (CD or DVDs).

When content is delivered through the content network, its descriptive information (metadata) is also delivered. The metadata information may be embedded within the media file(s) or it may be sent as separate data files. Some of the descriptive data may include text that is used for closed captioning compliance.

Figure 6.2 shows a contribution network that is used with an IPTV system. This example shows that programming that is gathered through a contribution network can come from a variety of sources that include satellite connections, leased lines, virtual networks, microwave links, mobile data, public data networks (e.g. Internet) and the use of stored media (tapes and DVDs).

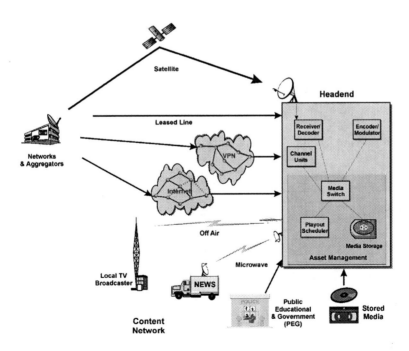

Figure 6.2, IPTV Contribution Network

## Program Transfer Scheduling

Program transfer scheduling is the set upand management of times and connection types that media will be transported to the IPTV system. IPTV systems have a limited amount of media storage for television programs so they typically schedule the transfer programming a short time (possibly several days) before it will be broadcasted in their system.

The cost of transferring content can vary based on the connection type (e.g. satellite versus Internet) and the data transfer speed. In general, the faster the data transfer speed, the higher the transfer cost. The scheduling of program transfer during low network capacity usage periods and at lower speed can result in significant reduction in transfer cost.

Figure 6.3 shows how an IPTV system may use transfer scheduling to obtain programs reliably and cost effectively. This example shows that the IPTV system may select multiple connection types and transfer speeds. This example shows that the selection can depend on the program type (live versus scheduled) and transfer cost.

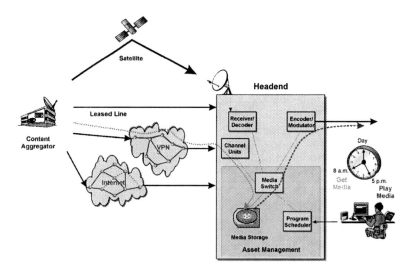

Figure 6.3, IPTV Program Transfer Scheduling

## Content Feeds

A content feed is a media source that comes from a content provider or stored media system. The types of content feeds that may be used in IPTV systems range from network feeds (popular programming) to video feeds from public events (government programming).

### Off Air Feed

An off air feed is a content source that comes from an antenna system that captures programming from broadcasted radio signals (off air radio transmission). The off air feed converts broadcasted radio channels into a format that can be retransmitted on another system (such as an IPTV system). Off-air feeds are used to retransmit locally broadcasted content on the IPTV system.

### Network Feed

A network feed is a media connection that is used to transfer media or programs from a network to a distributor of the media or programs.

### Local Feed

Local feed is a media connection that is used to transfer content from local sources. Examples of local feeds include connection from sportscasts, news crews and live studio cameras.

### Truck Feed

A truck feed is a media connection that is used to transfer content from mobile news vehicle source. Examples of truck feeds include cellular and microwave connections.

### Helicopter Feed

Helicopter feed is a media connection that is used to transfer content from airborne sources. Examples of helicopter feeds include microwave and private radio connections.

Live Feed

A live feed is a media connection that is used to transfer media or programs from a device that is capturing in real time (such as a mobile camera) to a distributor of the media or programs.

Government Access Channel

A government access channel is a media source that is dedicated to informing citizens of public related information. Examples of government programming include legal announcements, property zoning, public worker training programs, election coverage, health related disease controls and other public information that is related to citizens.

Educational Access Channel

An educational access channel is a media source that is dedicated to education. Educational programming may come from public or private schools. Examples of educational programming include student programming, school sporting events, distance learning classes, student artistic performances and the viewpoints and teachings of instructors.

Public Access Channel

A public access channel is a media source that is dedicated to allowing the public to create and provide programming to a broadcast system. Examples of public programming include local events and subjects that members of a community have an interest in.

Syndication Feeds

Syndication feeds are media connections or sources that are used to transfer media or programs from a syndicated network to a distributor of the media or programs. An example of a syndicated feed is really simple syndication (RSS) feed. An RSS feed provides content such as news stories via the Internet. RSS allows content from multiple sources to be more easily distributed. RSS content feeds are often commonly identified on web sites by an orange rectangular icon.

Emergency Alert System (EAS)

Emergency alert system is a system that coordinates the sending of messages to broadcast networks of cable networks, AM, FM, and TV broadcast stations; Low Power TV (LPTV) stations and other communications providers during public emergencies. When emergency alert signals are received, the transmission of broadcasting equipment is temporarily shifted to emergency alert messages.

Figure 6.4 shows some of the different types of content sources that may be gathered through a contribution network. This table shows that content sources include off-air local programs, entertainment from national networks, local programs, government access channels (public information), education access, public access (residents), syndication (shared sources), and the emergency alert systems.

| Content Sources | Notes |
| --- | --- |
| Off-air feed | Local programs |
| Network feed | Entertainment and nationwide programs |
| Local feed | Local information |
| Truck feed | Event and news information |
| Helicopter feed | Traffic and news |
| Live feed | Local news |
| Government access channel | Public information |
| Education access channel | Schools and learning sources |
| Public access channel | Community residents |
| Syndication feeds | News and shared resources |
| Emergency alert system | Public safety |

Figure 6.4, Contribution Network Programming Sources

# Headend

The headend is the master distribution center of an IPTV system where incoming television signals from video sources (e.g., DBS satellites, local studios, video players) are received, amplified, and re-modulated onto TV channels for transmission into the IPTV distribution system.

The incoming signals for headend systems include satellite receivers, off-air receivers and other types of transmission links. The signals are received (selected) and processed using channel decoders. Headends commonly use integrated receiver devices that combine multiple receiver, decoding and decryption functions into one assembly. After the headend receives, separates and converts incoming signals into new formats, the signals are selected and encoded so they can be retransmitted (or stored) in the IPTV network. These signals are encoded, converted into packets and sent into IPTV packet distribution system.

Figure 6.5 shows a diagram of a simple headend system. This diagram shows that the headend gathers programming sources, decodes, selects and retransmits video programming to the IP distribution network. The video sources to the headend typically include satellite signals, off air receivers, microwave connections and other video feed signals. The video sources are scrambled to prevent unauthorized viewing before being sent to the cable distribution system. The headend receives, decodes and decrypts these channels. This example shows that the programs that will be broadcasted are supplied to encoders produce IP television program streams. The programs are sent into the IPTV distribution network to distribution points (e.g. media servers) or directly to end users devices (e.g. set top boxes).

An IPTV system has expanded to include multiple regions including local headend locations, which may be distributed over a large geographic region. Local headends may be connected to regional headends and regional headends may be connected to a super headend. To reduce the cost of an IPTV system, headend systems can be shared by several distribution systems.

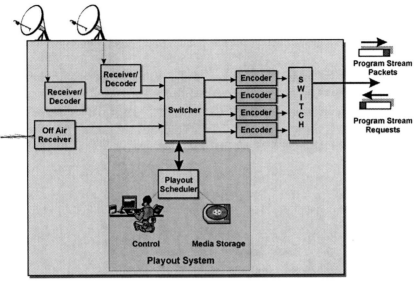

Figure 6.5, Headend System

## Integrated Receiver Decoder (IRD)

An integrated receiver and decoder is a device that can receive, decode, decrypt and convert broadcast signals (such as from a satellite system) into a form that can be transmitted or used by other devices.

In headend systems, IRDs are commonly used to demodulate and decrypt the multi-program transport stream (MPTS) from a satellite antenna. The IRD has a receiver that can select and demodulate a specific channel. The decoder divides an incoming channel into its component parts. A decryptor can convert the encrypted information into a form that can be used by the system. An interface converter may change the format of the media so that it can be used by other devices.

The inputs to an IRD (the front end) can include a satellite receiver or a data connection (such as an ATM or IP data connection). The types of processing that an IRD performs can vary from creating analog video signals to creating high definition video digital formats. The outputs of an IRD range from simple video outputs to high-speed IP data connections. Companies that produce IRDs commonly offer variations of IRD (such as analog and digital outputs) that meet the specific needs of the IPTV system operator.

Figure 6.6 shows the basic function of an integrated receiver device that is used in an IPTV system to receive satellite broadcasted signals and decode the channels. This example shows that an IRD contains a receiver section that can receive and demodulate the MPTS from the satellite. This IRD can decode and decrypt the MPTS to produce several MPEG digital television channels.

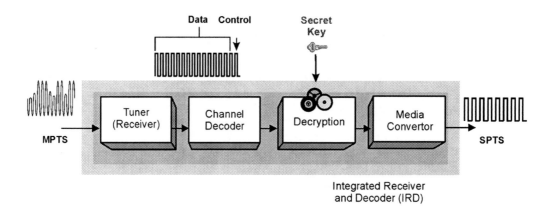

Figure 6.6, Headend Integrated Receiver and Decoder

## Off Air Receivers

An off air receiver is a device or assembly that can select, demodulate and decode a broadcasted radio channel. Broadcast receivers are used in cable television systems to receive local broadcasted channels so they can be re-broadcasted in the local cable television system.

In some countries (such as the United States), IPTV operators are required to rebroadcast local television channels on their cable television systems. These "must carry" regulations are government requirements that dictate that a broadcaster or other media service provider must retransmit (carry) or make available to carry a type of program or content.

Off-air receivers contain a tuner (receiver), demodulator and decoder for analog and/or digital television signals. The off-air receiver contain a tuning head that allows it to select (or to be programmed to select) a specific television channel. Off-air receivers may be simple analog television tuners (e.g. NTSC, PAL or SECAM) or they may be capable of demodulating and decoding digital television channels (e.g. DTT).

## Encoders

An encoder is a device that processes one or more input signals into a specified form for transmission and/or storage. A video encoder is a device used to form a single (composite) color signal from a set of component signals. An encoder is used whenever a composite output is required from a source (or recording) that is in a component format.

## Transcoders

A transcoder is a device or assembly that enables differently coded transmission systems to be interconnected with little or no loss in functionality.

## Packet Switch

A packet switch is a device in a data transmission network that receives and forwards packets of data. The packet switch receives the packet of data, reads its address, searches in its database for its forwarding address, and sends the packet toward its next destination.

A packet switch is used in an IPTV headend to select and forward packets from television program streams to their destination in the IPTV system. Their destination may be a media server or media hub that redistributes the packet stream to a group of recipients (viewers watching the same program) or to an individual viewing device (such as a viewer who is watching a video on-demand program).

## Asset Management

Asset management is the process of acquiring, maintaining, distributing and the elimination of assets. Assets for television systems are programs or media files. Assets are managed by workflow systems. Workflow management for television systems involves the content acquisition, metadata management, asset storage, playout scheduling, content processing, ad insertion and distribution control.

Content assets are acquired or created. Each asset is given an identification code and descriptive data (metadata) and the licensing usage terms and associated costs are associated with the asset. Assets are transferred into short term or long term storage systems that allow the programs to be retrieved when needed. Schedules (program bookings) are set upto retrieve the assets from storage shortly before they are to be broadcasted to viewers. When programs are broadcasted, they are converted (encoded) into forms that are suitable for transmission (such as on radio broadcast channels or to mobile telephones).

Figure 6.7 shows some of the common steps that occur in workflow management systems. This diagram shows that a workflow management system starts with gathering content and identifying its usage rights. The descriptive metadata for the programs is then managed and the programs are stored in either online (direct), nearline (short delay) or offline (long term) storage systems. Channel and program playout schedules are created and setup. As programs are transferred from storage systems, they may be processed and converted into other formats. Advertising messages may be inserted into the programs. The performance of the system is continually monitored as programs are transmitted `through the distribution systems.

Figure 6.7, Television Workflow Management Systems

## Content Acquisition

Content acquisition is the gathering content from networks, aggregators and other sources. After content is acquired (or during the content transfer), content is ingested (adapted and stored) into the asset management system.

Ingesting content is a process for which content is acquired (e.g. from a satellite downlink or a data connection) and loaded onto initial video servers (ingest servers). Once content is ingested it can be edited to add commercials, migrated to a playout server or played directly into the transmission chain.

Content acquisition commonly involves applying a complex set of content licensing requirements, restrictions and associated costs to the content. These licensing terms are included in content distribution agreements. Content licensing terms may define the specific type of systems (e.g. cable, Internet or mobile video), the geographic areas the content may be broadcasted (territories), the types of viewers (residential or commercial) and specific usage limitations (such as number of times a program can be broadcasted in a month). The content acquisition system is linked to a billing system to calculate the royalties and other costs for the media.

## Metadata Management

Metadata management is the process of identifying, describing and applying rules to the descriptive portions (metadata) of content assets. Metadata descriptions and formats can vary so metadata may be normalized. Metadata normalization is the adjustment of metadata elements into standard terms and formats to allow for more reliable organization, selection and presentation of program descriptive elements.

Metadata may be used to create or supplement the electronic programming guide (EPG). An EPG is an interface (portal) that allows a customer to preview and select from a possible list of available content media. EPGs can vary from simple program selection to interactive filters that dynamically allow the user to filter through program guides by theme, time period, or other criteria.

## Playout Scheduling

Playout scheduling is the process of setting up the event times to transfer media or programs to viewers or distributors of the media. A playout system is an equipment or application that can initiate, manage and terminate the gathering, transferring or streaming of media to users or distributors of the media at a predetermined time schedule or when specific criteria have been met.

Playout systems are used to select and assign programs (digital assets) to time slots on linear television channels. Playout systems are used to set upplaylists that can initiate automatic playout of media during scheduled interviews or alert operators to manually set upand start the playout of media programs (e.g. taps or DVDs).

Playout systems may be capable of selecting primary and secondary events. Primary events are the program that will be broadcasted and secondary events are media items that will be combined or used with the primary event. Examples of secondary events include logo insertion, text crawls (scrolling text), voice over (e.g. narrative audio clips) and special effects (such as a squeeze back).

Figure 6.8 shows the playout scheduling involves selecting programs and assigning playout times. This diagram shows a playout system that has multiple linear television channels and that events are set upto gather and playout media programs.

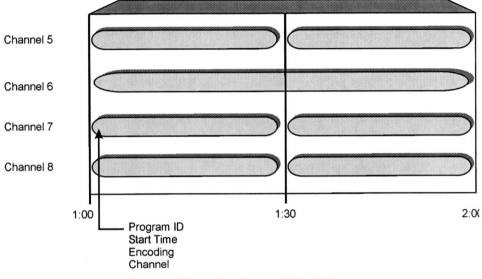

Figure 6.8, Television Playout Scheduling

Because the number of channels and programs is increasing, broadcasters may use playout automation to reduce the effort (workload) to set upplayout schedules. Playout automation is the process of using a system that has established rules or procedures that allows for the streaming or transferring media to a user or distributor of the media at a predetermined time, schedule or when specific criteria have been met (such as user registration and payment).

## Asset Storage

Asset storage is the maintaining of valuable and identifiable data or media (e.g. television program assets) in media storage devices and systems. Asset storage systems may use a combination of analog and digital storage media and these may be directly or indirectly accessible to the asset management system.

Asset management systems commonly use several types of storage devices that have varying access types, storage and transfer capabilities. Analog television storage systems may include tape cartridge (magnetic tape) storage. Digital storage systems include magnetic tape, removable and fixed disks and electronic memory.

Asset storage devices are commonly set upin a hierarchical structure to enable the coordination of storage media. Some of the different types of storage systems include cache storage (high speed immediate access), online storage, nearline storage, and offline storage.

Online storage is a device or system that stores data that is directly and immediately accessible by other devices or systems. Online storage types can vary from disk drives to electronic memory modules. Media may be moved from one type of online storage system to another type of online storage system (such as a disk drive) to another type of online storage (such as electronic memory) that would allow for rapid access and caching. Caching is a process by which information is moved to a temporary storage area to assist in the processing or future transfer of information to other parts of a processor or system.

Nearline storage is a device or system that stores data or information that is accessible with some connection set upprocesses and/or delays. The requirement to find and/or set upa connection to media or information on a nearline storage system is relatively small. Data or media that is scheduled to be transmitted (e.g. broadcasted) may be moved to nearline storage before it is moved to an online storage system.

Offline storage is a device or system that stores data or information that is not immediately accessible. Media in offline storage systems must be located and set upfor connection or transfer to be obtained. Examples of offline storage systems include storage tapes and removable disks.

## Content Processing

Content processing is adaptation, modification or merging of media into other formats. Content processing may include graphics processing, encoding and/or transcoding.

A graphics processor is an information processing device that is dedicated for the acquisition, analysis and manipulation of graphics images. Graphics processing may be required to integrate (merge or overlay) graphic images with the underlying programs.

Content encoding is the manipulation (coding) of information or data into another form. Content encoding may include media compression (reducing bandwidth), transmission coding (adapting for the transmission channel) and channel coding (adding control commands for specific channels).

Transcoding is the conversion of digital signals from one coding format to another. An example of transcoding is the conversion of MPEG-2 compressed signals into MPEG-4 AVC coded signals.

## Ad Insertion

Ad insertion is the process of inserting an advertising message into a media stream such as a television program. For broadcasting systems, Ad inserts are typically inserted on a national or geographic basis that is determined by the distribution network. For IP television systems, Ad inserts can be directed to specific users based on the viewer's profile.

An advertising splicer is a device that selects from two or more media program inputs to produce one media output. Ad splicers receive cueing messages (get ready) and splice commands (switch now) to identify when and which media programs will be spliced.

Cue tones are signals that are embedded within media or sent along with the media that indicates an action or event is about to happen. Cue tones can be a simple event signal or they can contain additional information about the event that is about to occur. An example of a cue tone is a signal in a television program that indicates that a time period for a commercial will occur and how long the time period will last.

Analog cue tone is an audio sequence (such as DTMF tones) that indicates a time period that will be available ('avail') for the insertion of another media program (e.g. a commercial).

An 'avail' is the time slot within which an advertisement is placed. Avail time periods usually are available in standard lengths of 10, 20, 30 or 40 seconds each. Through the use of addressable advertising, which may provide access to hundreds of thousands of ads with different time lengths, it is possible for many different advertisements, going to different audiences to share a single avail.

Digital program insertion is the process of splicing media segments or programs together. Because digital media is typically composed of key frames and difference pictures that compose a group of pictures (GOP), the splicing of digital media is more complex than the splicing of analog media that has picture information in each frame which allows direct frame to frame splicing.

Figure 6.9 shows how an ad insertion system works in an IPTV network. This diagram shows that the program media is received and a cue tone indicates the beginning of an advertising spot. When the incoming media is received by the splicer/remultiplexer, it informs the ad server that an advertising media clip is required. The ad server provides this media to the splicer which splices (attaches) each ad to the appropriate media stream. The resulting media stream with the new ad is sent to the viewers in the distribution system.

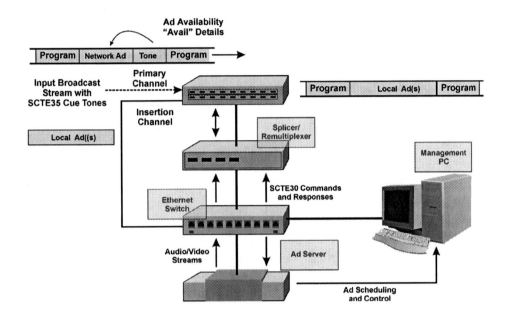

Figure 6.9, Television Ad Splicer

## Distribution Control

Distribution control is the processes that are used to route products or service uses to get from the manufacturer or supplier to the customer or end user. IPTV broadcasters may have several types of distribution networks including IP streaming, terrestrial radio broadcast systems (e.g. DTT), cable television distribution, mobile video and Internet streaming.

Distribution systems use a mix of media encoding formats that can include MPEG, VC-1 along with other compressed forms. The transmission of media to viewers ranges from broadcast (one to many), multicast (point to multipoint) and unicast (point to point).

Asset management systems use work orders to define, set upand manage assets. A work order is a record that contains information that defines and quantifies a process that is used in the production of media (e.g. television programs) or services. The development and management of assets is called workflow.

As the number of available programs and channels increases, it is desirable to automate the workflow process. Workflow automation is the process of using a system that has established rules or procedures that allows for the acquisition, creation, scheduling or transmission of content assets.

# Distribution Network

The distribution network is the part of an IPTV television system that connects the headend of the system (video and media sources) to the customer's location. IPTV signals are transmitted over a broadband IP network or a network that can transport broadband IP packets (such as an ATM or Ethernet system). IPTV distribution networks may be divided into a core network and an access network. IPTV systems may use a combination of individual channels (unicasting) or through shared channels (multicasting) in the distribution network.

Unicasting is the delivery media or data to only one client per communication session. If unicasting is used to provide broadcast services, a separate communication session must be established and managed between each user (client) and the broadcast provider (media server). If each user were connected directly to the headend of the system, the amount of data transmission required at the headend of the system would be substantial. Several thousand or hundreds of thousands of viewers would require 2 to 4 MByte connections each. To overcome this bandwidth challenge, popular programming (e.g. network programming) that is watched by multiple viewers at the same time is copied and redistributed (multicast) locally to the simultaneous viewers.

Multicast service is a one-to-many media delivery process that sends a single message or information transmission that contains a multicast address (code) that is shared by several devices (nodes) in a network. Each device that is part of a multicast group needs to connect to a router (node) in the network that is part of the multicast distribution tree. This means that the multicast media (such as an IPTV channel) is only sent to the users (viewers) who have requested it. The benefit of multicasting is the network infrastructure near the user (e.g. a home) only needs to provide one or two channels at once, drastically reducing the bandwidth requirements.

IPTV distribution systems commonly locate media servers throughout the distribution network to temporarily store popular television content. This allows for the supply of popular content (such as sports programs) during and shortly after the program has been broadcasted from local media servers.

Media servers can only hold a limited amount of programs. This means that programs that are stored in media servers located throughout the IPTV distribution network must be deleted after a short time to make storage memory space available for additional programs. For users who want to watch a program after it has been removed from a nearby server, their viewing request will be forward up towards a higher level sever (e.g. a regional hub) which may still contain the program. If the program is not available from the regional or central headend, the program may be requested and obtained from the content provider (e.g. the movie studio).

## Core Network

The core network is the central network portion of a communication system. The core network primarily provides interconnection and transfer between edge networks. Core networks in IPTV systems are commonly set upas fiber rings and spurs. A fiber ring is an optical network of network topology with a connection that provides a complete loop. The ring topology is used to provide a backup distribution path as traffic to be quickly rerouted in the other direction around the loop in the event of a fiber cut. A fiber spur is a fiber line that extends the fiber ring into another area for final distribution.

Figure 6.10 shows how an IPTV distribution system may use a combination of fiberoptic cable for the core distribution and broadband access for the local connection. This diagram shows that the multiple television channels at the headend of the IPTV television system are distributed through high-speed fiber cable. The fiber cable is connected in a loop around the local television service areas so that if a break in the optical cable occurs, the signal will automatically be available from the other part of the loop. The loop is connected (tapped) at regular points by a fiber hub that can distribute the optical signals to local access points. This example shows that the local access points may contain video servers so that popular content can be redistributed (shared between multiple viewers) or it will be available for viewing at a later time (time shifted television).

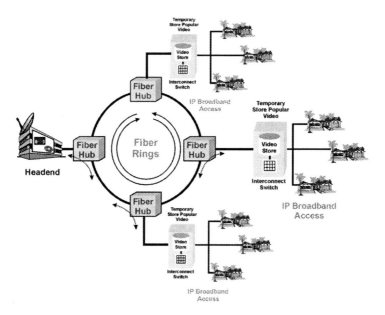

Figure 6.10, IP Television Program Distribution System

# Access Network

Access network is a portion of a communication network that allows individual subscribers or devices to connect to the core network. For IPTV systems, the end user device (e.g. set top box) communicates to the system using broadband data channels. The types of access networks that may be used by IPTV service providers include digital subscriber line (DSL), cable television (CATV), wireless broadband (WBB), powerline carrier (PLC) and optical networks.

Broadband access systems are networks that can transfer data signals at a rate of 1 Mbps or more. Broadband access systems that can provide data transmission rates of over 10 Mbps are called Ultra Broadband systems. Broadband access systems that may be suitable for IP television systems include digital subscriber line, cable modems, wireless broadband and powerline data.

Broadband access systems may be controlled (managed) or uncontrolled (unmanaged) systems. Broadband access systems that are controlled can guarantee the performance (data transmission rates). Broadband access systems that are uncontrolled offer best effort delivery.

If the broadband access system has data transmission rates that are several times the required data transmission rates for IP television (2 to 4 Mbps), unmanaged systems may provide acceptable data transmission rates for quality IP television programs.

## Digital Subscriber Line (DSL)

Digital subscriber lines transmit high-speed digital information, usually on a copper wire pair. Although the transmitted information is in digital form, the transmission medium is an analog carrier signal (or the combination of many analog carrier signals) that is modulated by the digital information signal.

DSL systems have data transmission rates that range from 1 Mbps to over 52 Mpbs. DSL systems have dedicated wire connections between the systems digital subscriber line access modem (DSLAM) and the user's DSL modem. Multiple DSL lines to the same location (multiple pairs of wires) may be combined to provide data transmission rates of over 100 Mbps (bonded DSL).

The data transmission rates in DSL systems can be different in the downlink and uplink directions (asymmetric) or they can be the same in both directions (symmetric). Because IP television data transmission is primarily from the system to the user (the downlink), asymmetric DSL systems (ADSL) are commonly used.

There are several types of DSL systems including asymmetric digital subscriber line (ADSL), symmetric digital subscriber line (SDSL) and very high bit rate digital subscriber line (VDSL). Of the different types of DSL, some have different versions with varying capabilities such as higher data transmission rates and longer transmission distances.

Because the telephone wires that DSL uses do not transfer high-frequency signals very well (high signal loss at higher frequencies), the maximum distance of DSL transmission is limited. In 2005, the typical maximum distance that DSL systems operated is 3 to 5 miles from the DSLAM. There is also a reduction in data transmission rate as the distance from the DSLAM increases. The further the distance the user is from the DSLAM, the lower the data transmission rate.

The data transmission required for each IP set top box in the home is approximately 2-4 Mbps. As the distance increases from the DSLAM to the customer, the number of set top boxes that can operate decreases. This means that customers who are located close to the DSLAM (may be located at the switching system) can have several set top boxes while IP television customers who are located at longer distances from the DSLAM may only be able to have one IP set top box. As the demand for IP television service increases, the DSL service provider can install DSLAMs at additional locations in their system.

Figure 6.11 shows how the number of simultaneous IP television users per household geographic serving area can vary based on the data transmission capability of the service provider. This example shows that each single IP television user typically requires 3 to 4 Mbps of data transfer. For a telephone system operator that uses distance sensitive DSL service, this example shows that the service provided will be limited to providing service to a single IP television when the data transfer rates are limited to 3-4 Mbps. When the data transfer rate is above 5-7 Mbps, up to 2 IP televisions can be simultaneously used and when the data transmission is between 10 to 14 Mbps, up to 3 IP televisions can be simultaneously used.

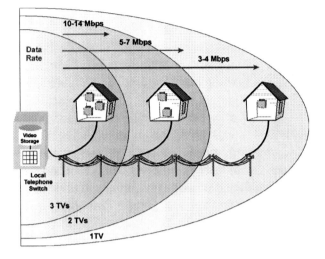

Figure 6.11, IP Television DSL Carrier Serving Area

## Cable Modem

A cable modem is a communication device that modulates and demodulates (MoDem) data signals through a cable television system. Cable modems select and decode high data-rate signals on the cable television system (CATV) into digital signals that are designated for a specific user.

Cable modems operate over the same frequency band as a television channel (6 MHz for NTSC and 8 MHz for PAL). Converting a television channel to cable modem provides data transmission rates of 30-40 Mbps per television channel from the system to the user and 2-4 Mbps from the user to the cable system. Each coaxial cable is capable of carrying over 100 channels. Cable modem users on a cable television system typically share a data channel.

Cable television systems have traditionally been used to transfer analog television channels from the media source (cable company headend) to end users (televisions). To provide cable modem service, some of the channels (usually the higher frequency channels) are converted from analog signals to digital video and data signals. A single analog television channel can be converted to approximately 6 digital television channels.

DOCSIS is a standard used by cable television systems for providing Internet data services to users. The DOCSIS standard was developed primarily by equipment manufacturers and CATV operators. It details most aspects of data over cable networks including physical layer (modulation types and data rates), medium access control (MAC), services and security.

Figure 6.12 shows a basic cable modem system that consists of a headend (television receivers and cable modem system), distribution lines with amplifiers, and cable modems that connect to customers' computers. This diagram shows that the cable television operator's headend system contains both analog and digital television channel transmitters that are connected to customers through the distribution lines. The distribution lines (fiber and/or coaxial cable) carry over 100 television RF channels. Some of the upper television RF channels are used for digital broadcast channels that transmit data to customers and the lower frequency channels are used to transmit digital information from the customer to the cable operator. Each of the upper digital channels can transfer 30 to 40 Mbps and each of the lower digital channels can transfer data at approximately 2 Mbps. The cable operator has replaced its one-way distribution amplifiers with precision (lin-

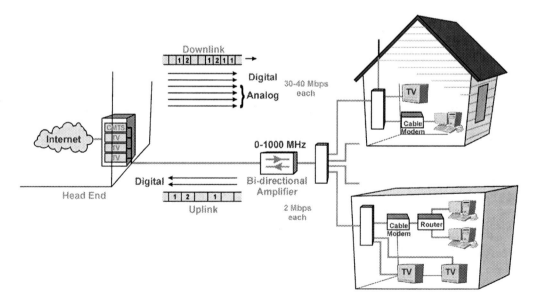

Figure 6.12, Cable Modem System

ear) high frequency bi-directional (two-way) amplifiers. Each high-speed Internet customer has a cable modem that can communicate with the cable modem termination system (CMTS) modem at the on-demand of the system where the CMTS system is connected to the Internet.

## Wireless Broadband

Wireless broadband is the transfer of high-speed data communications via a wireless connection. Wireless broadband often refers to data transmission rates of 1 Mbps or higher. Some of the available systems that can offer wireless broadband services include satellite systems, fixed microwave (MMDS/LMDS), 3G mobile communication, wireless LAN (WLAN) and WiMax.

While the installation time for wireless systems is relatively short, the cost of wireless system bandwidth has been traditionally higher than wired systems. This is because wireless systems generally cannot multiply their bandwidth by adding more wire or lines. Another typical tradeoff for a wireless system is higher mobility results in lower data transmission rates. Fixed wireless systems generally have higher data transmission rates than mobile wireless systems.

Satellite systems cover a very wide geographic area and they are well suited for one to many communications. The radio transmission from the digital broadcast satellite systems can provide data transmission at several hundred Mbps providing hundreds of digital video channels. While this system works well for simultaneous viewers, if the data transmission rate of each satellite channel is divided for number of viewers, the individual data rates will be relatively low (in the kbps). Some satellite systems use multiple beam transmission (spot beams) which can dramatically increase the available data transmission rates for each user.

Wireless broadband is commonly given the name "Wireless Cable" when it is used to deliver video, data and/or voice signals directly to end-users. Wireless cable provides video programming from a central location directly to homes via a small antenna that is mounted on the side of the house.

There are two basic types of wireless cable systems, Multichannel Multipoint Distribution Service (MMDS) and Local Multichannel Distribution Service (LMDS).

MMDS is a wireless cable service that is used to provide a series of channel groups, consisting of channels specifically allocated for wireless cable (the "commercial" channels). In the United States, MMDS service evolved from radio channels that were originally authorized for educational video distribution purposes. MMDS video broadcast systems have been in service since the early 1990's providing up to 33 channels of analog television over a frequency range from 2.1 to 2.69 GHz. Optionally, there are 31 "response channels" available near the upper end of the 2.5 to 2.69 band. These response channels were originally intended to transmit a voice channel from a classroom to a remote instructor.

The 3rd generation wireless requirements are defined in the International Mobile Telecommunications "IMT-2000" project developed by the International Telecommunication Union (ITU). The IMT-2000 project that defined requirements for high-speed data transmission, Internet Protocol (IP)-based services, global roaming, and multimedia communications. After many communication proposals were reviewed, two global systems are emerging; wideband code division multiple access (WCDMA) and CDMA2000/EVDO.

The maximum data transmission rates for 3G mobile communication systems can be over to 2 Mbps. However, the highest data transmission rates only can be achieved when the user is located close to the radio tower (cell site) and their interference levels (signals from other towers and users) are low. As a result, much of the live digital video transmission for mobile communication systems in 2005 had data transmission rates of approximately 144 kbps.

A wireless local area network (WLAN) is a wireless communication system that allows computers and workstations to communicate data with each other using radio waves as the transmission medium. The 802.11 industry standard and its various revisions are a particular form of Wireless LAN.

802.11 WLAN is commonly referred to as "Wi-Fi" (wireless fidelity). To help ensure Wi-Fi products perform correctly and are interoperable with each other, the Wi-Fi Alliance was created in 1999. The Wi-Fi Alliance is a nonprofit organization that certifies products conform to the industry specification and interoperates with each other. Wi-Fi® is a registered trademark of the Wi-Fi Alliance and the indication that the product is Wi-Fi Certified™ indicates products have been tested and should be interoperable with other products regardless of who manufactured the product.

Wireless LANs can be connected to a wired LAN as an extension to the system or it can be operated independently to provide the data connections between all the computers within a temporary ("ad-hoc") network. WLANs can be used in both indoor and outdoor environments.

Wireless LANs can provide almost all the functionality and high data-transmission rates offered by wired LANs, but without the physical constraints of the wire itself. Wireless LAN configurations range from temporary independent connections between two computers to managed data communication networks that interconnect to other data networks (such as the Internet). Data rates for WLAN systems typically vary from 1 Mbps to more than 50 Mbps.

Wireless LAN systems may be used to provide service to visiting users in specific areas (called "hot spots"). Hot spots are geographic regions or service access points that have a higher than average amount of usage. Examples of hot spots include wireless LAN (WLAN) access points in coffee shops, airports, and hotels.

There are several different versions of 802.11 WLAN systems that have evolved over time and some (not all) of these versions are compatible with other versions of 802.11 wireless LAN. The key differences include frequency band, type of wireless access, and maximum data transmission rates.

Figure 6.13 shows how 802.11 LAN systems have evolved over time. This diagram shows that the original 802.11 specification offered 1 or 2 Mbps data transmission rates and operated at 2.4 GHz. This standard evolved through new modulation to produce 802.11b that operated at 2.4 GHz and provided data transmission rates up to 11 Mbps. This diagram also shows that a new 802.11a system was developed that provides data transmission rates up to 54 Mbps at 5.7 GHz. To help provide high-speed data transmission rates and provide backward compatibility to 802.11 and 802.11b systems, the 802.11g systems was developed that offers 54 Mbps data transmission in the 2.4 GHz range.

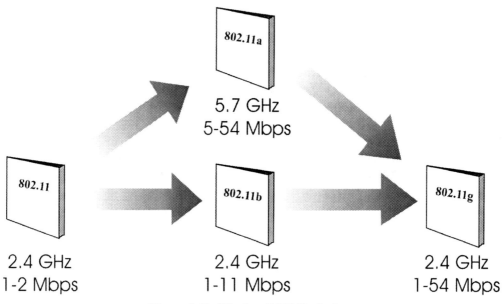

Figure 6.13, Wireless LAN Evolution

WiMax is a name for the IEEE 802.16A point to multipoint broadband wireless industry specification that provides up to 268 Mbps data transmission rate in either direction with a maximum transmission range of approximately 3-10 km [1]. The WiMax system is primarily defined to operate in either the 10 GHz to 66 GHz bands or the 2 GHz to 11 GHz bands.

Like the short range 802.11 wireless local area network (WLAN) specification, the 802.16 systems primarily defines the physical and media access control (MAC) layers. The system was designed to integrate well with other protocols including ATM, Ethernet, and Internet Protocol (IP) and it has the capability to differentiate and provide differing levels of quality of service (QoS).

Figure 6.14 shows the key components of a WiMax system. This diagram shows that the major component of a WiMax system include subscriber station (SS), a base station (BS) and interconnection gateways to datacom (e.g. Internet), telecom (e.g. PSTN) and television (e.g. IPTV).. An antenna and receiver (subscriber station) in the home or business converts the microwave radio signals into broadband data signals for distribution in the home. In this example, a WiMax system is being used to provide television and broadband data communication services. When used for television services, the WiMax system converts broadcast signals to a data format (such as IPTV) for distribution to IP set top boxes. When WiMax is used for broadband data, the WiMax system also connects the Internet through a gateway to the Internet. This example also shows that the WiMax system can reach distances of up to 50 km.

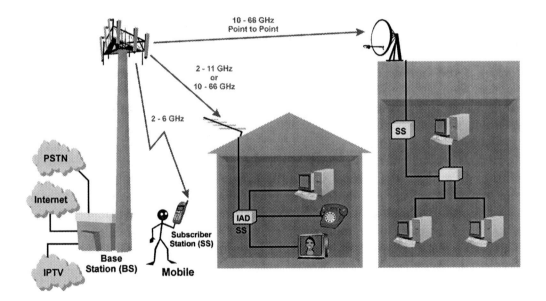

Figure 6.14, WiMax System

## Power Line Carrier (PLC)

Power line carrier is the simultaneous sending of a carrier wave information signal (typically a data signal) on electrical power lines to transfer information. A power line carrier signal is above the standard 60 Hz or 50 Hz frequency that transfers the electrical energy.

Power line carrier systems used in wide area distribution transfer signals from a data node located in the power system grid to a data termination box located in the home. Power line carrier systems typically use multiple transmission channels (low and medium frequencies) to communicate between the power system data node and the data termination box.

Because power lines were not designed for low and medium frequency signals, some of the signal energy leak from the power lines and some undesired signals are received into the power lines from outside sources (such as AM radio stations). Because the interference is typically different for each frequency, each power line transmission carrier signal is handled independently. If the signal energy is lost or if other signals interfere with one frequency, one or several other transmission channel frequencies can be unused. Multiple transmission channels can be combined to provide high-speed data transmission rates.

Powerline transformers typically block high frequency data signals so data routers or bridges are typically installed near transformers. Utility companies typically set updata nodes near transformers and each data node provides signals to homes that are connected to the transformers. There are often many homes per transformer (over 1,000 homes) in developing countries and a lesser number of homes per transformer (less then 12 homes) in developed countries.

Figure 6.15 shows three types of communication systems that use an electrical power distribution system to simultaneously carry information signals along with electrical power signals. In this example, the high voltage portion of the transmission system is modified to include communication transceivers that can withstand the high voltages while coupling/transferring their information to other receivers that are connected to the high voltage lines. This type of communication could be used to monitor and control power distribution equipment such as relays and transformers or as a high-speed data backbone transmission connection using fiber. This example also shows a power line distribution system that locates a communication node (radio or fiber hub) near a transformer and provides a data signal to homes connected to a transformer. This system could allow customers to obtain Internet access or digital telephone service by plugging the computer or special telephone into a standard power socket. The diagram also shows how a consumer may use the electrical wiring in their home as a distribution system for data (e.g. Ethernet) communication.

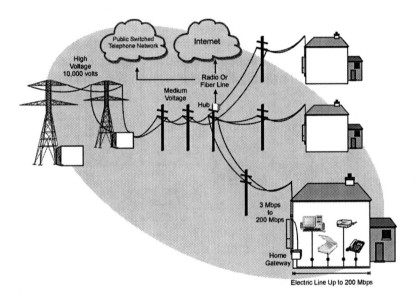

Figure 6.15, Powerline Broadband Data Transmission

## Optical Access Network

An optical access network is a portion of a communication network that allows individual subscribers or devices to connect to the core network through optical connections. Optical access networks may use fiber to connect to a node (distribution point) in a neighborhood, to a node outside a home or to a optical network termination (ONT) that is located in the home or business.

Figure 6.16 shows an example of an optical access network. This diagram shows that the telecommunications network is converted to a format suitable for an optical access network by an optical line termination (OLT). For a fiber to the neighborhood (FTTN) system, the optical signal travels to an optical network unit (ONU) located in a geographic region close to the customer (their neighborhood). The ONU converts the optical signal into a format that is sent to a network termination (NT) at the customer's premises. For a fiber to the curb (FTTC) system, the optical signal is sent to an ONU located at the curb near the customer's premises. For a fiber to the home (FTTH) system, the optical signal travels to an optical network termination (ONT) that is located in the customer's premises. The ONT functions as the optical signal converter and the network end termination point.

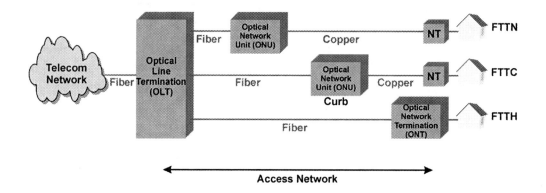

Figure 6.16, Optical Access Network

## Premises Distribution

A premises distribution network is the equipment and software that is used to transfer data and other media in a customer's facility, home or personal area. A PDN is used to connect terminals (computers) to other networks and to wide area network connections. Some of the common types of PDN are wired Ethernet, Wireless LAN, Powerline, Coaxial and Phoneline Data.

PDN networking systems have transitioned from low speed data, simple command and control systems to high-speed multimedia networks along with the ability to transfer a variety of media types that have different transmission and management requirements. Each of the applications that operate through a PDN can have different communication requirements that typically include a maximum data transmission rate, continuous or bursty transmission, packet delay and jitter and error tolerance. The PDN system may manage these connections using a mix of protocols that can define and manage quality of service (QoS). Transmission medium types for premises distribution include wired Ethernet (data cable), wireless, power-line, phoneline and coaxial cables.

Wired LAN systems use cables to connect routers and communication devices. These cables can be composed of twisted pairs of wires or optical fibers. Wired LAN data transmission rates vary from 10 Mbps to more than 1 Gbps. While wired Ethernet systems offer high data throughput and reliability, many homes do not have dedicated wiring installed for Ethernet LAN networks and for the homes that do have data networks, the outlets are often located near computers rather than near televisions.

Wireless local area network (WLAN) systems allow computers and workstations to communicate with each other using radio propagation as the transmission medium. The wireless LAN can be connected to an existing wired LAN as an extension, or it can form the basis of a new network. Wi-Fi television distribution is important because it is an easy and efficient way to get digital multimedia information where you need it without adding new wires. Wireless LAN data transmission rates vary from 2 Mbps to over 54 Mbps and higher data transmission rates are possible through the use of channel bonding. WLAN networks were not designed specifically for multimedia. In the mid 2000s, several new WLAN standards were created to enable and ensure different types of quality of service (QoS) over WLAN.

Power line carrier systems allow signals to be simultaneously transmitted on electrical power lines. A power line carrier signal is transmitted above the standard 60 Hz power line power frequency (50 Hz in Europe). Power line premises distribution for television is important because televisions, set-top boxes, digital media adapters (DMAs) and other media devices are already connected to power outlets already installed in a home or small businesses. Older (legacy) power line communication systems had challenges with wiring systems that used two or more phases of electrical power. Today, with the benefit of modern signal processing techniques and algorithms, most of these impairments no longer are an impediment to performance and some powerline data systems have data transmission rates of over 200 Mbps.

Coaxial cable premises distribution systems transfer user information over coaxial television lines in a home or building. Coaxial distribution systems may simply distribute (split) the signal to other televisions in the home or they may be more sophisticated home data networks. When coax systems are set upas simple distribution systems, they are set upas a tree distribution system. The root of the tree is usually at the entrance point to the home or building. The tree may divide several times as it progresses from the root to each television outlet through the use of signal splitters. When coaxial systems are set upas data networks, data signals at high frequencies (above 860 MHz) are combined with broadcast signals over the same coaxial lines. Coaxial cable data transmission rates vary from 1 Mbps to over 1 Gbps and many homes have existing cable television networks and the outlets which are located near video accessory and television viewing points.

Telephone wiring premises distribution systems transfer user information over existing telephone lines in a home or building. Because telephone lines may contain analog voice signals and data signals (e.g. DSL), premises distribution on telephone lines uses frequency bands above 1 MHz to avoid interference with existing telephone line signals. Telephone data transmission rates vary from 1 Mbps to over 300 Mbps. There are typically several telephone line outlets installed in a home and they may be located near television viewing points, making it easy to connect television-viewing devices.

Figure 6.17 shows the common types of premises distribution systems that can be used for IP television systems. This diagram shows that an IP television signal arrives at the premises at a broadband modem. The broadband modem is connected to a router that can distribute the media signals to forward data packets to different devices within the home such as IP televisions. This example shows that routers may be able to forward packets through power lines, telephone lines, coaxial lines, data cables or even via wireless signals to adapters that receive the packets and recreate them into a form the IP television can use.

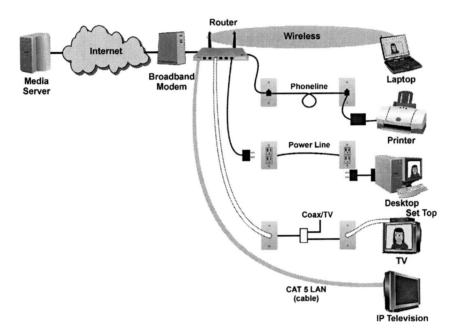

Figure 6.17, IPTV Television Premises Distribution

References:

1. "What is Wi-Max?" WiMax Forum, www. wimaxforum.org.

# Chapter 7

# Premises Distribution Networks (PDN)

A premises distribution network is the equipment and software that is used to transfer data and other media in a customer's facility, home or personal area. A PDN is used for media devices such as digital televisions, computers and audio visual equipment to each other and to other networks through the use of wide area network connections. PDN systems may use wired Ethernet, Wireless LAN, Powerline, Coaxial or Phoneline connections.

PDN networking systems have transitioned from low speed data, simple command and control systems to high-speed multimedia networks. Premises distribution systems have evolved to include the ability to transfer a variety of media types that have different transmission and management requirements. Each of the applications that operate through a PDN can have different communication requirements which typically includes a maximum data transmission rate, continuous or bursty transmission, packet delay, jitter and error tolerance. The PDN system may manage these connections using a mix of protocols that can define and manage quality of service (QoS).

Figure 7.1 shows the common types of premises distribution systems that can be used for IP television systems. This diagram shows that an IP television signal arrives at the premises at a broadband modem. The broadband modem is connected to a router that can distribute the media signals to forward data packets to different devices within the home such as IP televisions. This example shows that routers may be able to forward packets through power lines, telephone lines, coaxial lines, data cables or even via wireless signals to adapters that receive the packets and recreate them into a form the IP television can use.

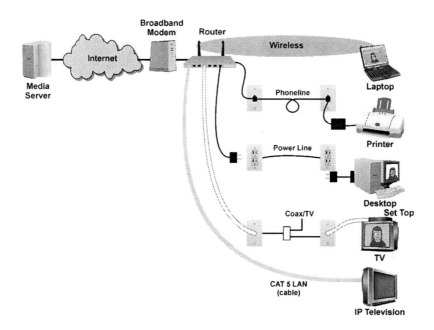

*Figure 7.1, Premises Distribution Systems*

## Home Multimedia Service Needs

Multimedia networking is the process of transferring multiple forms of media (e.g. digital audio, data and digital video) through a communication network (such as a home network). Multimedia networking is used to connect computers, media players and other media sources and players to each other and to send and receive media from wide area network connections.

The communication requirements for multimedia networks in the home is based on the applications that are used in the home along with how much and the times that the applications and services will be used. The typical types of applications that will be used in the home include telephone, Internet access, television, interactive video and media streaming.

## Telephone

Telephony is the use of electrical, optical, and/or radio signals to transmit sound to remote locations. Generally, the term telephony means interactive communications over a distance. Traditionally, telephony has related to the telecommunications infrastructure designed and built by private or government-operated telephone companies.

Digital telephone systems (such as Internet telephones) may use a PDN to connect within a home. Uncompressed digital telephone signals have a data transmission of approximately 90 kbps. The communication must be real time with minimal transmission delays. Some transmission errors are acceptable. Home telephone usage is typically is less than 1 hour and the usage time (call duration) only lasts for a few minutes.

## Internet Access

Internet access is the ability for a user or device to connect to the Internet. Internet access can be used for web browsing or data transfer applications.

Internet communication tends to require high-speed data transmission in short bursts. Short delays are acceptable and transmission errors are usually automatically overcome by Internet protocols. In general, the amount of data transmission for Internet connections continues to increase as rich media sources (video streaming) increase.

## Television

A television service is the transmission of television program material (typically video combined with audio). Digital television signals require high-speed data transmission rates (2-20 Mbps) that are used for several hours per day. For some types of digital compression such as MPEG-2, even with small amounts of transmission errors (0.1%) can result in television signals that are unwatchable.

A single home can have multiple televisions which require multiple simultaneous data streams. High definition television channels are becoming more available which is likely to result in an increase in the required bandwidth for television signals. Television viewing often occurs at the same time (e.g. evening hours).

## Interactive Video

Interactive video services are processes that allow users to interactively provide and receive audio and/or visual information. The interactive parts range from simple text messages to the simultaneous transfer of large data or media files.

Interactive video services such as video conferencing requires a mid-level amount of data transmission (approximately 200kbps) for medium periods of time (minutes to hours).

## Media Streaming

Media streaming applications include security camera connections or music distribution (e.g. home theater) systems. Although the amount of data transmission for these may be low, they may be used for long durations.

The data transmission requirements for home media networks may be increased as a result of multiple connection paths (e.g. double or triple hops) and as a result of playback modes that can require multiple streams (such as trick modes).

A double hop is a transmission path that is routed through a communication network where it takes two connection paths to reach its destination rather than through a single direct path. Double hop connections through a home media network can use up to two times as much data transmission capacity.

Trick mode is a media player function that allows for variations in the presentation of media. Examples of trick mode are fast forward, rewind and pause. Trick mode commands enable remote control features for the playout of media. During trick play mode, additional streams may be created. Some of the trick mode features that may increase data transmission requirements include fast sliding and slow looking. Fast slide is the rapid acceleration of playing (presenting) media (such as video) through the rapid movement (dragging) of a position indicator on a media slide control. Show looking is the rapid browsing through a media program (such as a video) by dragging the media position indicator through a portion or over the entire length of a media program (fast slide).

Figure 7.2 shows some of the types of communication devices used in a home and their estimated data transmission requirements. This table shows that some devices may require connections through a gateway to other networks (such as to the Internet or to television systems). This table also shows that the highest consumption of bandwidth occurs from television channels, especially when simultaneous HDTV channels are access on multiple television sets. This table suggests that a total home media capacity of 70 Mbps to 100 Mbps is required to simultaneously provide for all devices within a home and that a residential gateway have broadband capability of 50+ Mbps.

| Service | Bandwidth | Number of Devices | Bandwidth Residential Gateway |
|---|---|---|---|
| TV | 2 to 20 Mbps | 3 | 2 to 54 Mbps |
| Digital Video Recorder (DVR) | 2 to 20 Mbps | 1 | 0 |
| Home Theater | 1 to 6 Mbps * Audio | 1 System | 0 |
| Internet Browsing | 1 to 2 Mbps | 1 to 5 | 1 to 10 Mbps |
| Printer | 0.5 to 1 Mbps | 1 to 5 | 0 |
| Digital Imaging | 1 to 20 Mbps | 1 to 3 | 0 |
| Digital Telephone | 0.2 Mbps | 1 to 5 | 0.2 to 1 Mbps |
| Online Gaming | 0.2 to 1 Mbps | 1-3 | 0.2 to 3 Mbps |
| Video Capturing | 0.1 to 1 Mbps | 1-10 * Security Cameras | 0 |
| Portable Audio | 0.1 to 20 Mbps | 1 to 3 | 0 |
| Total | 70 Mbps to 100 Mbps | | 2 to 60 Mbps+ |

Figure 7.2, Home Multimedia Bandwidth Requirements

# Home Multimedia System Needs

The requirements for home multimedia systems vary depending on the needs of the viewers, the needs of the operator and existing home network systems. Some of the general factors that are considered when selecting a home multimedia networking system include co-existence, no new wiring requirements, data rates, quality of service (QoS), security, cost and installation.

## Co-Existence

Co-existence is the ability of a device or system to operate near or with another device and/or system. Premises distribution systems typically co-exist with one or more types of systems including wireless LAN, CATV, Satellite, Home Audio/Visual and others. The PDN system may share the same medium (cables and/or radio waves).

To co-exist, these systems may use segmentation of medium where a portion of the medium (e.g. frequency range) is used by one system and another segment (e.g. different frequency range) is used by the PDN system.

## Intersystem Interference

Intersystem interference is the interaction of signals or data in one system with signals or data in another system where the interaction results in the performance degradation or unwanted changes in the operation of either system. PDN systems may experience or cause interference with other systems that may be located at adjacent locations (e.g. neighbors home). Intersystem interference may occur between systems that are physically connected or it may occur by signal leakage. When interference occurs, it may result in distortion or reduced transmission capacity.

Interference between wired systems may be reduced or eliminated by the installation of a point of entry (POE) filter. A POE filter is a device (analog or digital) that is designed to allow or block specific signals from entering or leaving a building or facility (at the point of entry).

## Protocols

Protocols are the languages, processes, and procedures that perform functions used to send control messages and coordinate the transfer of data. Protocols define the format, timing, sequence, and error checking used on a network or computing system. While there are several different protocol languages used for PDN services, the underlying processes are commonly based on Ethernet and IP protocols. Protocols for distribution services (such as RTP) simply ride on top (are encapsulated) into the data packets that

## No New Wires

"No New Wires" is a concept that can be used when adding additional communication capabilities in a home or building without adding any new wireless. The NNW concept does allow for the use of existing wires as part of the new communication system.

## Data Rates

Data rate is the amount of information that is transferred over a transmission medium over a specific period of time. The required data transmission rates for home media networks can be characterized by average and peak requirements.

The average data transmission rates is decreased as a result or improved media compression capability (e.g. moving from MPEG-2 to MPEG-4) but it is increased by the addition of new applications (media portability and media conversion).

Peak when multiple users select options that increate bandwidth such as trick play or rapid channel changing. These features if performed simultaneously by all users in the system could increase the data transmission requirements by 1.5x to 2x.

## Quality of Service (QoS)

Quality of service is one or more measurement of desired performance and priorities of a communications system. Some of the important QoS characteristics for IPTV PDN systems include error rates, transmission delay and jitter.

### *Error Rates*

Error rate is a ratio between the amount of information received in error as compared to the total amount of information that is received over a period of time. Error rate may be expressed in the number of bits that are received in error of the number of blocks of data (packets) that are lost over a period of time.

Some digital video formats (such as MPEG-2) are more sensitive to packet error rates (burst errors) than packet errors. Acceptable error rates for video distribution is a packet error rate (PER) of $10^{-6}$ and a bit error rate (BER) of $10^{-9}$ [1].

## *Transmission Delay*

Transmission delay is the time that is required for transmission of a signal or packet of data from entry into a transmission system (e.g. transmission line or network) to its exit of the system. Common causes of transmission delay include transmission time through a transmission line (less than the speed of light), channel coding delays, switching delays, queuing delays waiting for available transmission channel time slots, and channel decoding delays.

The overall transmission delay for video signals is an issue for real-time communication (e.g. video conferencing) but not much of an issue for broadcast television or on-demand services.

## *Jitter*

Jitter is a small, rapid variation in arrival time of a substantially periodic pulse waveform resulting typically from fluctuations in the wave speed (or delay time) in the transmission medium such as wire, cable or optical fiber. Digital video and audio is generally more sensitive to jitter rather than to overall transmission delay. Jitter can cause packets to be received out of sequence and miss their presentation window.

## Home Coverage

Home coverage is the available area (for wireless) or number of outlets (for wired systems) over which the signal strength of a premises distribution network is sufficient to transmit and receive information, media or data.

## Security

Security is the ability of a person, system or service to maintain its desired well being or operation without damage, theft or compromise of its resources from unwanted people or events. Security for IPTV home multimedia distribution systems includes privacy and rights management.

## Privacy

Privacy is a level of protection that is minimal, corresponding to a moderate amount of effort on the part of the eavesdropper to understand the private communication, but not so good as the better levels designated by the words "secret" or "secure." Most PDN systems have the capability of encrypting media so other devices or neighboring systems cannot decode.

## Rights Management

Rights management is a process of organization, access control and assignment of authorized uses (rights) of content. Rights management may involve the control of physical access to information, identity validation (authentication), service authorization, and media protection (encryption). Rights management systems are typically incorporated or integrated with other systems such as content management system, billing systems, and royalty management.

## Cost

Cost is the fees our resources that are required to produce or provide a product or service. PDN costs can include equipment, marketing, license or franchise fees, transaction and service support costs.

## Installation

Installation is the locating and configuration of equipment and/or wiring either inside or outside of buildings or facilities. When the equipment has passed operational and performance tests installation may be complete.

While it is possible for customers to self-install premises distribution systems, in the initial launch phase of IPTV, many carriers have chosen to have their workers or contractors to install the equipment to ensure that both the PDN and IPTV signals are operating correctly.

## *Peer to Peer Operation*

Peer to peer distribution is the process of directly transferring information, services or products between users or devices that operate on the same hierarchical level (usage types that have similar characteristics).

Many PDN systems are designed for plug and play where the devices use self discovery. Self discovery is the processes used to request and receive information that is necessary for a device to begin operating. Self discovery may involve finding devices within a system that can provide it's the identification address, name, and services of other devices.

## *Upgradeable*

Upgradability is the ability of a device or systems to be modified, changed or use newer components and/or as technology innovations become available. PDN systems may be upgradeable through the use of software downloads.

## *Remediation*

Remediation is the correction of a condition or problem. For PDN systems, remediation may involve the dispatching of a qualified technician or field service representative to validate, configure or correct faulty installations.

## *Remote Diagnostics*

Remote diagnostics is a process and/or a program that is built into a device or network component that allows users or devices from remote locations to access information or control operations that can be used to discover the functionality or performance of a device or system connection.

# Home Networking Technologies

Home networking systems use key technologies to allow them to reliably provide broadband transmission. These technologies include adaptive modulation, synchronized transmission, interference avoidance, power level control and channel bonding.

## Adaptive Modulation

Adaptive modulation is the process of dynamically adjusting the modulation type of a communication channel based on specific criteria (e.g. interference or data transmission rate). Adaptive modulation allows a transmission system to change or adapt as the transmission characteristics change. For some PDN systems, the characteristics of the transmission line can dramatically change over short periods of time.

PDN systems typically share transmission resources (such as a power line) and the characteristics of these transmission lines can dramatically change (e.g. light switch). Adaptive modulation allows the PDN transmission system to adapt to the new conditions that occur such as slowing the transmission rate when interference occurs and to return to a more efficient transmission system when the characteristics of the transmission line changes again.

## Echo Control

Echo control is the processes that are used to identify echoes to transmitted signals and the processes that are used to remove the distortions caused by the signal echoes.

For wired PDN systems, there can be many connection paths and some of these connection paths may not be properly installed or terminated which can result in signal reflections. There can also be changes in the PDN systems that occur from the installation, control or removal devices or configurations. These endpoints that are not properly terminated can cause signal reflections and the signal reflections can cause signal distortion.

Signal reflection is the sharp changing of the direction of a transmitted signal as it passes from one transmission medium to another (transmission channel or device). When the characteristics of the mediums are different (impedance), a reflected signal will be generated. Some of the energy of the forward signal (incident signal) is redirected (reflected) back towards the signal source. Echoed signals may be received at other devices and these cause distortion in their received signals.

The distortion caused by echoed signals in PDN systems can usually be removed through the use of echo canceling. Echo cancellation is a process of extracting an original transmitted signal from the received signal that contains one or more delayed signals (copies of the original signal). Echo canceling may be removed by performing via advanced signal analysis and filtering.

## Synchronized Transmission

Synchronized transmission is the transferring of information during a time period that is previously defined (time synchronized) or transmission occurs after another event (such as a synchronization message). Synchronized transmission is used to ensure (guarantee) that certain types of information will be provided at specific times when it is required.

The use of synchronized transmission reserves transmission resources (bandwidth reservation) for time critical information such as television or telephone media. The use of synchronized transmission provided for reduced transmission delays (limited latency) minimizes jitter and provides for low error rates.

PDN systems may use one of the devices that are connected to the PDN as a master controller. The master controller coordinates transmissions within the PDN. Devices request transmission resources (e.g. reserved bandwidth). If the master unit has the bandwidth available and it determines the requesting device is authorized to use the service, it reserves the bandwidth in messages it periodically sends throughout the PDN.

To perform synchronized transmission, the master unit periodically sends media access plan (MAP) messages that has a list of devices and their assigned transmission schedules and priorities for media transmission. After a PDN device has decoded the MAP message and has determined that its time to transmit has occurred, it can begin transmitting without coordinating with other device (contention free).

## Interference Avoidance

Interference avoidance is a process that adapts the access channel sharing method so that the transmission does not occur on specific frequency bandwidths. By using interference avoidance, devices that operate within the same frequency band and within the same physical area can detect the presence of each other and adjust their communication system to reduce the amount of overlap (interference) caused by each other. This reduced level of interference increases the amount of successful transmissions therefore increasing the overall efficiency and increased overall data transmission rate.

## Power Level Control

Power level control is a process of adjusting the power level of a transmitter on a communication link to minimize the amount of energy that is transmitted while ensuring enough signal level is transmitted so a desired reception quality can be achieved. In general, the closer the transmitter is located to the receiver, the lower the amount of power level that needs to be transmitted.

Power control is typically accomplished by sensing the received signal strength level and relaying power control messages between the receiver and transmitter that indicates or instructs the transmitter to increase or decrease it's output power level.

## Channel Bonding

Channel bonding is the process of combining two or more communication channels to form a new communication channel that can use and manage the combined capacity of the bonded transmission channels. Channel bonding can be used on Wireless PDN systems to increase the overall data transmission rate and to increase the reliability of transmission during interference conditions.

Figure 7.3 shows how multiple transmission channels can be bonded together to produce a single channel with higher data transmission rates. This diagram shows how two transmission channels can be combined using a bonding protocol. This diagram shows that a bonded session is requested and negotiated on a single communication channel. Once the bonded session has been setup, the bonding protocol is used to monitor and manage the bonded connection.

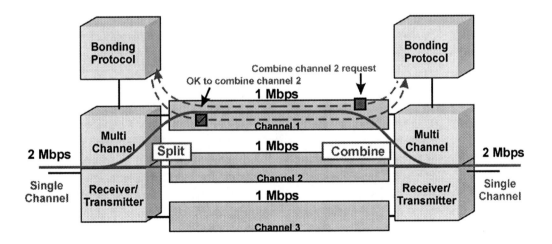

*Figure 7.3, Channel Bonding*

# Transmission Types

Transmission medium types for premises distribution include wired Ethernet (data cable), wireless, powerline, phoneline and coaxial cables.

## Wired LAN

Wired LAN systems use cables to connect routers and communication devices. These cables can be composed of twisted pairs of wires or optical fibers. Wired LAN data transmission rates vary from 10 Mbps to more than 1 Gbps.

Wired LAN systems are typically installed as a star network. The star point (the center of thee network) is usually a router or hub that is located near a broadband modem. LAN wiring is not a commonly installed in many homes and when LAN wiring is installed,

LAN connection outlets are unlikely to be located near television viewing points. When wired LAN systems use twisted cable, the data transmission rate (cable rating) is based on the number of twists in the cable. The higher the number of twists, the higher the maximum data transmission rate and the higher the category rating of the LAN cable.

A data cable that is used for wired LAN networks is classified by the amount of data it can carry in by the Electronics Industry Association/Telecommunications Industry Association EIA/TIA 586 cabling standard. In general, category 1 rated cable is unshielded twisted-pair (UTP) telephone cable (not suitable for high-speed data transmissions). Category 2 cable is UTP cable that can be used for transmission speed up to 4Mbps. Category 3 UTP cable can be used at data transmission speeds of 10Mbps. Category 4 UTP cable can transmit up to 16Mbps and is commonly used for Token Ring networks. Category 5 cable is rated for data transmission speeds up to 100Mbps. Category 5E (enhanced) has the same frequency range as Category 5 with a lower amount of signal transfer (crosstalk) so it can be used for 1 Gbps Ethernet systems. Category 6 cable has a frequency rating of 250 MHz.

## Wireless

A wireless local area network (WLAN) allows computers and workstations to communicate with each other using radio propagation as the transmission medium. The wireless LAN can be connected to an existing wired LAN as an extension, or it can form the basis of a new network. While adaptable to both indoor and outdoor environments, wireless LANs are especially suited to indoor locations such as office buildings, manufacturing floors, hospitals and universities.

Wi-Fi distribution of IPTV streams from broadband providers is the use of unlicensed radio frequencies to distribute multimedia signals to IP-enabled receiving devices in a home or nearby location.

Figure 7.4 shows how an in-home Wi-Fi system can be used for IP Television (IPTV) premises (home) distribution. This diagram shows that a broadband modem is installed in the home that has WLAN with premises distribution capability. This example shows that the broadband modem is located at a point that is relatively far from other devices in the home. The broadband modem is connected to a wireless access point (AP) that retransmits the broadband data to different devices through the home including a laptop computer, Wi-Fi television and an IP set top box (IP STB) that has a built-in Wi-Fi receiver.

Wi-Fi distribution is important because it is an easy and efficient way to get digital multimedia information where you need it without adding new wires. Some consumers have refused to add new IPTV services due to rewiring or having to retrofit their homes to support it.

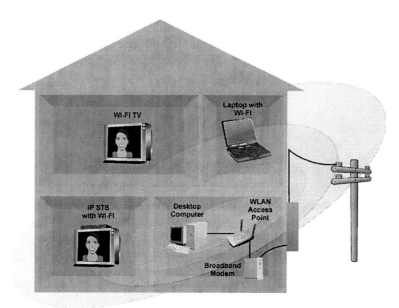

*Figure 7.4, Wireless IPTV Distribution*

Wireless LAN premises distribution systems transfer user information over a WLAN system in a home or building. Wireless LAN data transmission rates vary from 2 Mbps to over 54 Mbps. Higher data transmission rates are possible through the use of channel bonding.

Multimedia signals such as television and music are converted into WLAN (Ethernet) packet data format and distributed through the home or business by wireless signals. Some versions of the 802.11 WLAN specifications include the ability to apply a quality of service (QoS) to the distributed signals giving priority to ensure that time sensitive information (such as video and audio) can get through and non-time sensitive information (such as web browsing).

Figure 7.5 shows the different product groups of 802.11 systems and how the data transmission rates have increased in the 802.11 WLAN over time as new more advanced modulation technologies are used. The first systems could only transmit at 1 Mbps. The current evolution of 802.11A and 802.11G allows for data transmission rates of up to 54 Mbps.

| | 802.11 1 | 802.11 2 | 802.11A 3 | 802.11B 4 | 802.11G 5 |
|---|---|---|---|---|---|
| Access Method | FHSS | DSSS | DSSS | DSSS | DSSS |
| Modulation Type | GFSK | DBPSK DQPSK | OFDM QPSK QAM | CCK | OFDM QPSK QAM |
| Frequency | 2.4 Ghz | 2.4 Ghz | 5.7 Ghz | 2.4 Ghz | 2.4 Ghz |
| Data Transmission Rate | 1-2Mbps | 1-2Mbps | 9-54Mbps | 5.5-11Mbps | 9-54Mbps |

*Figure 7.5, Wireless LAN Standards*

WLAN networks were not designed specifically for multimedia. In the mid 2000s, several new WLAN standards were created to enable and ensure different types of quality of service (QoS) over WLAN.

## Power Line

Power line carrier is a signal that can be simultaneously transmitted on electrical power lines. A power line carrier signal is transmitted above the standard 60 Hz powerline power frequency (50 Hz in Europe).

The power line communication systems developed in the 1970s used relatively low frequencies such as 450 kHz to transfer data on power lines. The amount of data that could be transferred was limited and the applications typically consisted of controlling devices such as light switches and outlets. Some of the early home automation systems include: X-10, CeBus and LONworks.

Power line communication signals travel from their entry onto the home power line system (typically from a powerline-enabled broadband modem such as WiMAX, DSL, cable modem or a fiber optic Network Interface

Device in the home) through a variety of conductive powerline pathways until it reaches a destination device.

The percentage of outlets or electrical devices within a home that can receive and process power line data signals is referred to as a statistically derived "home coverage" figure. The power line communications coverage within a home is expressed as a rate of speed over a certain percentage of outlets. Therefore, as an example, 25 Mbps at 80% coverage should be interpreted as 80% of the power outlets in a house will provide at least 25 Mbps in bandwidth. This is different than an average performance figure. The home coverage figure is important to a service provider, as it defines a relative throughput performance index within an average home for some percentage of total outlets. A number of factors influence overall performance, including the distance the signals have to travel (size of home), the size of the wire, and the number of undetermined outlets.

Older (legacy) power line communication systems had challenges with wiring systems that used two or more phases of electrical power. They were especially sensitive to noise impairments like brush motors, halogen lamps and dimmer switches. Today, with the benefit of modern signal processing techniques and algorithms, most of these impairments no longer are an impediment to performance. Whereas in the past, a powerline communications device only provided connectivity to 70-80% of the outlets reliably, connectivity to any outlet in a home is a virtual certainty.

Looking at the early implementations of powerline communications systems, much of the coverage challenge was due to the difficulty of powerline communication signals to propagate on electrical powerlines that are on different phases. The amount of signal energy that transfers across power lines that are connected to different phases of a home's power grid is called cross phase coupling. To help the power line communication signal to cross over different phases of an electrical system, cross phase couplers could be installed. While some of these cross phase couplers could be installed by the homeowner at outlets that had both phases (e.g. a dryer outlet), cross phase couplers improved the home coverage ratio, but the improvements were not always enough to solve the overall challenges of connectivity and as a result, operation of these home automation systems were not always reliable.

Figure 7.6 shows how existing power line distribution systems in a home (such as X-10) can distribute data signals in a home. This diagram shows that medium voltage electricity (6,000 to 16,000 Volts) is supplied to a step down transformer near the home. This transformer produces two or three phases of power ranging from 110 V to 240 V. These electrical signals pass through an electrical distribution panel (circuit breaker panel) in the home to supply outlets and lights. This diagram shows an X-10 power line communication system where an X-10 switch in the wall controls an X-10 outlet

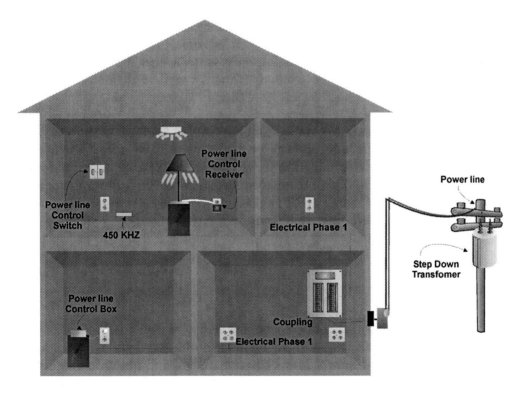

*Figure 7.6, Power Line Communication Control System*

receiver by sending control signals over the power line on the same circuit (same phase) to allow a user to control a lamp. This example also shows an X-10 power line control box plugged into another outlet on a different circuit (different phase) and the control signal must cross over from one phase to the other phase (cross over in the electrical panel) so it can reach the X-10 outlet receiver.

Although power line communication systems could technically transfer data in the 1970s and 1980s, improvements in the power line data transfer rates necessary for home data networking did not occur until the early 2000s. Part of the motivation to make advances in power line communication was the increased need for home networking.

Power line premises distribution for IPTV is important because televisions, Set-Top Boxes, Digital Media Adapters (DMAs) and other media devices are already connected to power outlets already installed in a home or small businesses. These lines can be used to transmit rich multimedia content where it is desired.

Intelligent powerline distribution systems are able to adapt to the varying characteristics of the in-home power grid and should be able to transfer media between any device enabled to "hear" the content being distributed on the powerline. Powerline distribution systems convert the different types of media (Internet data, Voice-over-IP packets, or IP video streams) into information that is modulated onto high frequency carriers imposed on the powerline.

## Coaxial

Coaxial cable premises distribution systems transfer user information over existing coaxial television lines in a home or building. Coaxial cable data transmission rates vary from 1 Mbps to over 1 Gbps and many homes have existing cable television networks and the outlets which are located near video accessory and television viewing points.

Coaxial systems are commonly setup as a tree distribution system. The root of the tree is usually at the entrance point to the home or building. The tree may divide several times as it progresses from the root to each television outlet through the use of signal splitters.

In addition to the signal energy that is lost through the splitters, some signal energy is lost through attenuation on the coax cable itself. The type of coaxial cable, the signal frequency and the length of the cable determine the amount of energy that is lost. As the length of coaxial lines increases and the number of ports (splitters) increases, the amount of attenuation also increases.

Coax lines to and from the cable television (CATV) company may contain analog and digital television and cable modem signals. Cable television distribution systems use lower frequencies for uplink data signals and upper frequencies for downlink data and digital television signals. Some of the center frequencies are used for analog television signals. These frequency bands typically range up to 850 MHz. Coax premises distribution systems can use frequencies above the 850 MHz frequency band to transfer signals to cable television jacks throughout the house. Adapter boxes or integrated communication circuits convert the video and/or data signals to high frequency channels that are distributed to different devices located throughout the house. To ensure the coax premises distribution signals do not transfer out of the home to other nearby homes, a blocking filter may be installed.

Figure 7.7 shows how an in-home coaxial cable distribution system is typically used to distribute television signals in a home. This diagram shows that a cable connection is made at a demarcation ("demarc") point on the outside of a home or building. This cable connects to a signal splitter (and optionally an amplifier) that divides the signal and sends it to several locations throughout the home. This example shows that the coaxial systems may use lower frequency bands for return connections (data modems), middle frequency bands for analog television channels, and upper frequency bands for digital television channels.

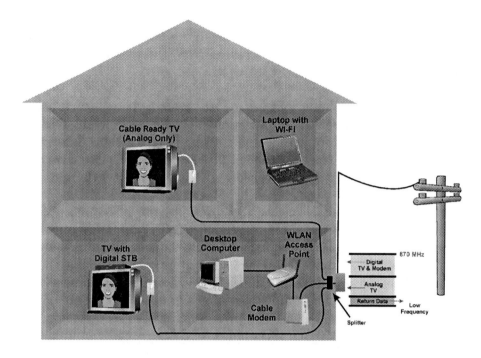

*Figure 7.7, Coax Television Distribution system*

Smart coaxial premises distribution for IPTV is important because television and other media devices are already connected to a coaxial line that is already installed in a home or business. These lines can be used to get multimedia information where you need it.

Smart coaxial distribution systems are usually able to distribute or interconnect media between several devices simultaneously. Smart coaxial systems modulate different media (Internet data, voice-over-IP packets, or IP video streams) into different portions of the frequency spectrum on the coaxial lines.

Because the coax cable is shielded and RF channels are virtually free from the effects of interfering signals, coaxial cable provides a large information pipe that is capable of distributing multiple wide radio frequency channels. Coax is already installed in many homes, and television outlets are commonly located near media equipment such as televisions, VCRs and cable modems. Coaxial cable is easy to install and expand.

To overcome some of the loss in a cable distribution system, RF amplifiers are used. If these RF amplifiers are used, they can introduce distortion and may block signals from two-way operation. If amplifiers are needed to strengthen the RF signal, they should be installed before the TVnet master device so as to prevent disruption of the IP network.

## Phoneline

Telephone wiring premises distribution systems transfer user information over existing telephone lines in a home or building. Telephone data transmission rates vary from 1 Mbps to over 300 Mbps. Telephone line outlets may be located near television viewing points making it easy to connect IP television viewing devices.

Telephone lines to and from the telephone company may contain analog voice signals, uplink data signals, and downlink data signals. These frequency bands typically range up to 1 MHz (some DSL systems go up to 12 MHz). Phone line premises distribution systems may use frequencies above the 1 MHz frequency band to transfer signals to telephone jacks throughout the house. Adapter boxes or integrated communication circuits convert the video and/or data signals to high frequency channels that are distributed to different devices located throughout the house. To ensure the phone line signals do not transfer out of the home to other nearby homes, a blocking filter may be installed.

The earliest telephone line home multimedia communication systems were used to allow computers to transfer files with each other (home data network) and to connect data communication accessories such as printers. These early telephone line data communication systems sent a limited amount of data using frequencies slightly above the audio frequency band. These early systems had relatively low data transmission rates and they were fairly limited when compared to automatic telephone line communication systems.

Phoneline data transmission involves converting data or information signals into a form (such as a modulated carrier frequency) that can travel along a communication line. As the signals travel down the telephone lines, a portion of the signal is lost through the wires (absorbed or radiated). Signal frequency, the type of wire, how the wire is installed and the length of the wire are key factors in determining the amount of energy that is lost. Generally, as the length of a telephone increases and the number of outlets increases, the amount of attenuation also increases.

A challenge for transmitting data over the phoneline is the presence of interfering signals and the variability of the characteristics of the telephone lines. Interference signals include telephone signals (ringing, DTMF, modems) and signals from outside source (such as AM radio stations).

Phoneline data communication systems need to work through existing telephone lines (which may be unshielded) that are connected with each other in a variety of ways (e.g. looped or spliced throughout a house) and may have a variety of telephone devices and accessories attached to it. Variability can be caused by poor installation of telephone wiring, telephone cords and changes in the characteristics of the telephone devices and accessories. To overcome the effects of interference and the variability of transmission lines, phoneline system use adaptive transmission systems.

Because some of the energy leaks out of the telephone line, the maximum signal level authorized by regulatory authorities (such as the Federal Communication Commission) is relatively low.

Figure 7.8 shows how existing phoneline distribution systems in a home or business can be used to distribute both voice and data signals in a home or building. This diagram shows that an incoming telephone line is directly connected to all telephone outlets in the home or building. In some cases, the lines are spliced together and in other cases, the lines are simply connected in sequence. This example shows that a phoneline data network uses frequencies above the telephone line audio to distribute the data signals by sharing the phoneline.

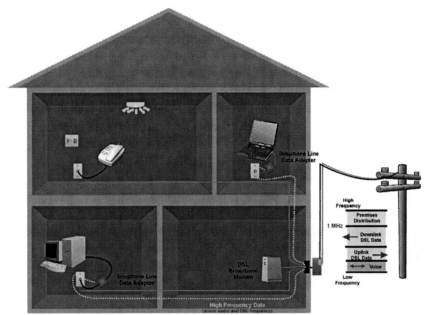

*Figure 7.8, Phoneline Communication System*

# Premises Distribution Systems

Premises distribution systems are the combination of equipment, protocols and transmission lines that are used to distribute communication services in a home or building. Premises distribution systems may be used on one or several types of transmission lines such as coax, twisted pair, powerline or wireless. Some of the more popular types of premises distribution systems used for IPTV systems include HomePlug™, HomePNA™, TVNet, MoCA and 802.11 WLAN.

## HomePlug™

HomePlug is a system specification that defines the signals and operation for data and entertainment services that can be provided through electric power lines that are installed in homes and businesses. Development of the HomePlug specification is overseen by the HomePlug® Powerline Alliance.

### HomePlug 1.0

HomePlug 1.0 is a powerline data communication system that transfers data through existing power lines at 14 Mbps using higher frequencies (from 4 MHz to 21 MHz). The HomePlug system uses a higher frequency range than previous powerline data communication systems which enables signals to couple phases in the electrical distribution panel. As a result, the coverage of HomePlug power line communication systems approached 99% without the need for cross phase couplers or professional installers.

As the signals travel down the power lines, a portion of the signal is lost through the wires (absorbed or radiated). Signal frequency, the type of wire, how the wire is installed and the length of the wire are key factors in determining the amount of energy that is lost. Generally, as the length of a power line increases and the number of outlets increases, the amount of attenuation also increases. Signal loss at high frequencies through power lines can reach 50 dB and the dynamic range of HomePlug devices can be between 70 dB to 80 dB.

Because it is possible for HomePlug signals to travel on power lines shared between several homes, the signals are encrypted to keep the information private. The HomePlug 1.0 system encrypts (scrambles) information using 56-bit DES security coding. Only devices that share the same HomePlug encryption codes can share the information transferred between the devices.

One of the significant challenges for power line communication systems is the "sources and effects" of interference signals that can distort power line communication signals. Interference signals include motor noise, signal reflections, radio interference, changes in electrical circuit characteristics, variability in the amount of coupling across different phases of electrical circuits and stray transmission. The HomePlug system was designed to overcome these types of interference and in some cases, the HomePlug system can take advantage of them.

Motor noise is the unwanted emissions of electrical signals produced by the rapidly changing characteristics of a motor assembly. In most homes, motors are in a variety of appliances and they may be used at different locations at any time. The HomePlug system can adapt in real time to the distortions caused by motors and appliances.

Signal reflection is the changing of direction that a signal travels as it passes from one transmission medium to another (transmission channel or device). When the characteristics of the media are different (impedance), a reflected signal is generated. Some of the energy of the forward signal (incident signal) is redirected (reflected) back towards the signal source. When high-frequency HomePlug signals reach the ends of power lines, some of their signal energy is reflected back towards the transmitter. These reflected signals combine with the forward (incident) signals producing distortion. The HomePlug system includes sophisticated analysis of the signal and it is possible to use the signal reflections as an advantage rather than a challenge.

Some of the frequencies used by HomePlug systems are the same radio frequencies used by citizen band (CB) radios or AM broadcast radio stations. The HomePlug system divides the frequency band into many independently controlled sub-channels (using a modulation scheme known as Orthogonal Frequency Division Multiplexing – OFDM) so that when interference is detected (such as from a hair dryer), the sub-channels that are affected can be shut off. Because there are so many available sub-channels, this has little effect on the overall capacity of the HomePlug system.

Another common challenge with home power line communication is the dynamic change that can occur in electrical circuit characteristics as users use light switches and plug into or remove electrical devices from outlets. The HomePlug system is smart enough to sense and adjust for the signal channels due to changes in electrical circuit characteristics.

Each device that is a part of an in-home HomePlug network requires a HomePlug adapter or a HomePlug converter built into the device. The HomePlug adapters convert information signals (digital data) into frequency carriers that travel down the power line. The adapters also coordinate access to the power line communication system by first listening to ensure there is no existing activity before transmitting and stopping transmission when they detect information packet collisions have occurred.

Figure 7.9 shows how the HomePlug 1.0 system allows a power line distribution system to transfer data between devices connected to electrical outlets in a home. This example shows several computers operating in a home and communicating between other data communication devices plugged into outlets. This example shows that a computer can send a data signal to a printer located at an outlet in another room. This diagram shows that the high frequency signals from the computer travel down the electrical line to the electrical panel. Then at the electrical panel to the power line communication signal, it jumps across the circuit breakers and travels down the electrical lines to reach the HomePlug 1.0 adapter connected to a printer. This diagram also shows that the computer can send data signals down the power line to a HomePlug router connected to a broadband modem allowing the computer to connect to the Internet.

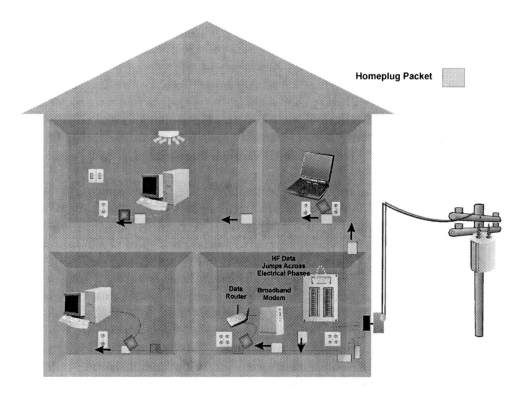

Figure 7.9, HomePlug 1.0 Distribution System

## *HomePlug Audio Visual (HomePlug AV)*

The HomePlug Audio Visual (HomePlug AV) specification was ratified in 2005 by the HomePlug Board of Directors, to provide home media networking. HomePlug AV was designed to give priority to media that requires time sensitive delivery (such as IPTV) while allowing reliable data communication (such as web browsing) to simultaneously occur. HomePlug AV uses a mix of random access (unscheduled) and reserved access (scheduled) data transfer. The carrier sense multiple access with collision avoidance (CSMA/CA) protocol, which provides for efficient transfer of bursty data while the scheduled TDMA system ensures real time media (such as digital video and audio) will be delivered without delays and will take priority on the wire over CSMA/CA traffic.

The HomePlug AV system is designed to be a 200 Mbps-class PHY device. Actual application layer performance will be less than that, as MAC efficiency needs to be contemplated in overall performance measurements. HomePlug AV uses a significantly higher spectrum (from 2 MHz to 28 MHz) which allows for good cross phase coupling because these higher frequencies can jump across circuit breakers to increase the signal availability in the home. Like HomePlug 1.0, the coverage of HomePlug AV power line communication systems approaches 99% without the need for cross phase couplers or professional installers.

The HomePlug AV system uses a more secure 128-bit AES encryption process to keep information private. The HomePlug AV system has better control of transmission delays (latency and jitter) and is designed to co-exist with the HomePlug 1.0 system. Because it operates at high frequencies, it does not interfere with power line control systems such as X-10.

The HomePlug AV system uses nearly 1000 separate narrowband carriers. The system can selectively shut off these narrowband carriers (frequency notches) when it senses interference.

Like the HomePlug 1.0 system, each platform that will be part of the HomePlug AV network requires a HomePlug AV adapter or a HomePlug AV transceiver built into the device. The adapters coordinate access to the power line communication system. This coordination involves either reserving time periods for transmission and/or dynamically coordinating access by listening to the signals on the power line to ensure there is no existing activity before they transmit and stopping transmission when they detect data transmission collisions have occurred.

Figure 7.10 shows how a HomePlug AV system can allow a power line distribution system to dynamically interconnect devices by connecting devices to electrical outlets and using high frequency signals to transfer information. This example shows that a HomePlug adapter coordinates the transmission of signals without interfering with the transmission between other devices. In this diagram, an IPTV service provider is sending a movie through a broadband modem to a television in the home. This example shows that the HomePlug system uses a HomePlug AV router to receive the signals from the broadband modem and route the information through the power lines to the television in the home using dedicated time slots to

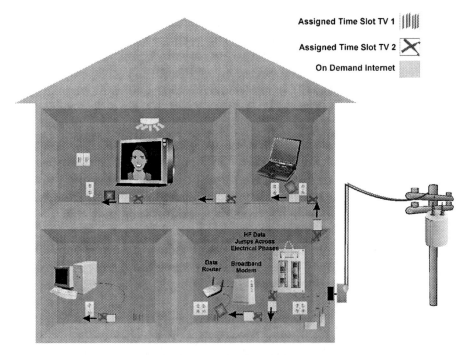

*Figure 7.10, HomePlug AV Distribution System*

ensure a reliable quality of service (QoS). This example shows that simultaneously, a computer is connected through a HomePlug AV adapter to the in-home broadband modem allowing it to connect to the Internet and transfer data on an "as available" basis.

It is possible for customers to self-install a HomePlug AV system. The system is setup to automatically detect (discover) other devices in the network. While most HomePlug devices are pre-configured with the most common settings, it is possible to customize the configuration of the HomePlug adapter devices by connecting them to a local computer and allowing the user to change key settings such as IP addresses and passwords.

Each device connected to a HomePlug system must have an Ethernet or USB port, which allows a connection to an adapter or have a HomePlug AV transceiver embedded within a platform. The adapter can be an external box or it can be designed into a product (such as a CD player). Adapters can be simple conversion devices (Ethernet to HomePlug) or they can include some packet transfer capability (routers or bridges). A HomePlug Ethernet bridge allows packets from a HomePlug network to enter into an Ethernet network (such as allowing data to the Internet).

HomePlug devices include nodes and bridges. Nodes simply convert data signals into the frequency signals that travel on the power lines. HomePlug bridges allow data packets to cross over into devices or other networks without any need to use configuration utilities.

The HomePlug AV powerline technology theoretical PHY transmission rate is 200 Mbps. There is a difference between physical data transmission rates (the "PHY" rate) and actual data throughput. The throughput is affected by the amount of overhead (control) data and the amount of information that is lost via interference and the access control process. Even with the overhead control information, the HomePlug AV system can have data transmission rates in excess of 100 Mbps. HomePlug AV technology is designed to co-exist with a variety of other systems including HomePlug 1.0, HomePlug 1.0 with Turbo, X-10, CeBus and LONworks.

The typical range of a HomePlug signal is approximately 1000 feet (300 meters). HomePlug technology generally works in most homes. A broad-based product ecosystem is rapidly forming for HomePlug AV technology. This ecosystem includes set-top boxes, routers, gateways, switches, displays, TVs, DMAs and entertainment electronics platforms. HomePlug devices that are certified to operate under the HomePlug specification will interoperate with each other.

## HomePNA™

The HomePNA is a non-profit association that works to help develop and promote unified information about a home multimedia network technologies, products and services. HomePNA initially started by defining home data networks over telephone lines and it has expanded its focus to multimedia distribution over telephone and coaxial lines.

Sending broadband information over the same wires used to transfer telephone lines and coaxial lines involves converting the information into signals that can be transferred down these existing wires without interfering with other existing signals.

Phoneline data communication systems use adapters to convert information from digital format to signals that can travel along the telephone lines. Phoneline data communication systems need to work through existing telephone lines (which may be unshielded) that are connected with each other in a variety of ways (e.g. looped or spliced throughout a house) and may have a variety of telephone devices and accessories attached to it.

The HomePNA system has now evolved into a high-speed home multimedia communication network that can transfer digital audio, video and data over telephone lines or coaxial lines. The HomePNA system has been designed to co-exist with both telephone signals (lower frequency range) and digital subscriber line signals. Digital subscriber line (DSL) systems typically occupy frequency bands from the audio range up to approximately 2 MHz.

To connect to the HomePNA network, each device must have an adapter or bridge to convert the information or data into a HomePNA signal. Adapters or bridges can be separate devices or they can be integrated into a device such as set top box. The adapters convert standard connection types such as wired Ethernet (802.3), universal serial bus (USB), Firewire (IEEE1394) and others.

A gateway or bridge may be used to link the HomePNA network to other systems or networks (such as the Internet). The HomePNA system was designed to allow a gateway (such as a residential gateway - RG) to be located anywhere in the home.

A challenge for transmitting data over the phonelines is the presence of interfering signals and the variability of the characteristics of the telephone lines. Interference signals include telephone signals (ringing, DTMF, modems) and signals from outside source (such as AM radio stations). Variability can be caused by poor installation of telephone wiring, telephone cords and changes in the characteristics of the telephone devices and accessories. To overcome the effects of interference and the variability of transmission lines, the HomePNA system uses an adaptive transmission system.

The HomePNA system monitors its performance and it can change the transmission characteristics on a packet by packet basis as the performance changes. The changes can include changing the modulation type and/or data transmission rates to ensure communication between devices connected to the HomePNA system is reliable.

## HomePNA 1.0

HomePNA 1.0 was the first industry specification that allowed for the use of data transmission over telephone lines. The HomePNA 1.0 system has a data transmission rate of 1 Mbps with a net throughput of approximately 650 kbps. HomePNA 1.0 systems are very low cost and are used throughout the world. They are popular in multi-tenant units in the far east.

## HomePNA 2.0

HomePNA 2.0 is a home networking system that is designed to provide medium speed data transmission rates and to allow priority to the transmission of different types of media. The HomePNA 2.0 system has a data transmission rate of 16 Mbps with a net data transmission throughput of approximately 8 Mbps. The HomePNA 2.0 system added new capabilities to home networking from the previous HomePNA 1.0 system including prioritized transmission for different types of media and an optional form of modulation to help ensure more reliable transmission during interference conditions.

Transmission prioritization of was added to the HomePNA system to allow for differentiation and prioritization of packets for time sensitive services. These prioritized services included IP telephone and video services where longer packet transmission delays were not acceptable.

The HomePNA 2.0 system included the ability to transmit data using a modulation scheme known as Frequency Diversity Quadrature Amplitude Modulation (FDQAM). When interference is detected (such as from an AM radio station), the modulation type can change from QAM to FDQAM increasing the robustness (reliability) of the signal. The FDQAM performs spectral multiplication where the same signal is transmitted on different frequencies (frequency diversity).

## HomePNA 3.0

HomePNA 3.0 is a multimedia home networking system that is designed to provide high-speed data transmission rates and to control the quality of service for different types of media. The HomePNA 3.0 system has a data transmission rate of 240 Mbps with a net data transmission throughput of approximately 200 Mbps and it is backward compatible with HomePNA 2.0.

HomePNA 3.0 was standardized in June 2003 and commercial products were available starting in May 2004. The industry standard has been adopted by the ITU as specification number G.9954.

HomePNA 3.0 uses either quadrature amplitude modulation (QAM) for high-speed transmission or frequency diverse quadrature amplitude modulation (FDQAM). Depending on the quality of the transmission line and the media requirements. The data transmission rate is constantly adjusted on a packet by packet basis by varying the symbol rates (increasing and decreasing the amount of shifts per second) and the number (precision) of decision points per symbol to ensure data can be received in varying conditions.

The HomePNA 3.0 system added the ability to the mix of controlled (contention free) and on demand (random access) transmission to home networking. The HomePNA 3.0 system accomplishes this by assigning one of the HomePNA devices in the system as a master control unit. The Master unit coordinates the transmission time periods of all of the HomePNA devices in the system through the periodic transmission of a media access plan message. A media access plan is a list of devices and their assigned transmission schedules and priorities for media transmission.

The HomePNA 3.0 system can provider for a mix of synchronized (guaranteed) and random access (unscheduled) transmission through the use of a master controller. The master coordinates all of the bandwidth reservation and transmission assignment and all the other units are slaves that follow the lead of the master.

The master can be selected automatically or it can be manually setup or programmed as the master. For example, an IPTV service provider can setup the residential gateway (broadband modem) to be the master unit of the HomePNA system.

Each HomePNA 3.0 device is typically capable of acting as a master or a slave. If the master unit is shut off or if it becomes disabled, the HomePNA system will automatically setup another unit as the master.

Figure 7.11 shows a HomePNA 3.0 system. This diagram shows that the master unit periodically sends MAP message and that each MAP message identifies which devices are assigned synchronized transmission and when unscheduled transmissions may occur. This example shows that scheduled devices only transmit during their assigned schedules and other devices compete for access during the remainder (unscheduled portion) of the media cycle.

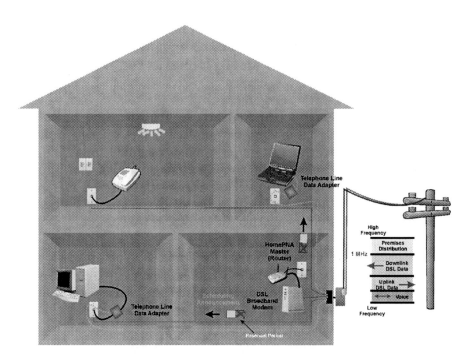

*Figure 7.11, HomePNA 3.0 Distribution System*

## HomePNA 3.1

HomePNA 3.1 is a multimedia home networking system that is designed to provide high-speed data transmission rates and to control the quality of service for different types of media over both telephone lines and coaxial cables. The HomePNA 3.1 system has a data transmission rate of 320 Mbps with a net data transmission throughput of approximately 250 Mbps. The HomePNA 3.1 system is backward compatible with HomePNA 3.0 and HomePNA 2.0

The HomePNA 3.1 system uses the same frequency band for the telephone line and coaxial line. This frequency band is above digital subscriber line signals and is below television channels. When operating on the coaxial line, the HomePNA 3.1 system only uses the scheduled mode which can be more than 90% efficient increasing the data rate available for the users. Because the coaxial line is a shielded line, the HomePNA system can increase its power (called "power boost"), allowing the effective distance the HomePNA system can operate to over 5,000 feet (1,500 meters).

Figure 7.12 shows how a HomePNA 3.1 system can use both telephone line and coaxial lines to be used. The HomePNA system allows devices to directly communicate with each other. This diagram shows that the HomePNA system can be used to distribute voice, data and video signals over both telephone lines and coaxial lines allowing it to be available at more locations in the home.

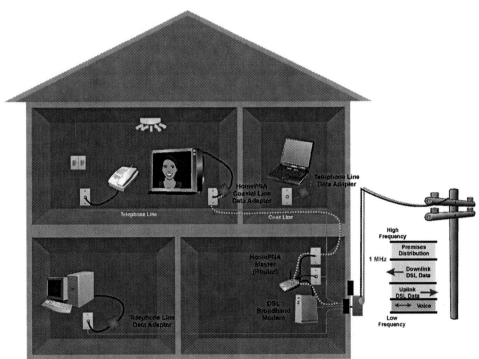

*Figure 7.12, HomePNA 3.0 Phoneline and Coax Distribution*

Figure 7.13 shows a summary of the different types of HomePNA systems and their capabilities. This table shows that the HomePNA system has evolved from a low speed data only to a high-speed multimedia multiple line type home networking system.

The typical range of a HomePNA signal is approximately 1000 feet (300 meters) on telephone lines. HomePNA devices that are certified to operate under the HomePNA specification will interoperate with each other.

| System | Data Rate | Type of Media | Notes |
|--------|-----------|---------------|-------|
| HomePNA 1.0 | 1 Mbps, (650 kbps throughput) | Data Only | QAM Only |
| HomePNA 2.0 | 16 Mbps (8 Mbps throughput) | Prioritized for Multimedia | FDQAM and QAM |
| HomePNA 3.0 | 240 Mbps (200 Mbps throughput) | Scheduled Guaranteed QoS for Digital TV | FDQAM and QAM |
| HomePNA 3.1 | 320 Mbps (250+ Mbps throughput) | Scheduled Guaranteed QoS for Digital TV | Telephone line and Coax |

*Figure 7.13, HomePNA Capabilities*

Some telephone devices and accessories that are lower quality can create interference with the HomePNA system. While this interference would typically only reduce the performance (e.g. data transmission rate), installing microfilters to block the unwanted interference from the telephone devices can solve it.

## Digital Home Standard (DHS)

Digital home standard is a system specification that defines the signals and operation for data and entertainment services that can be provided through electric power lines that are installed in homes and businesses. Development of the DHS specification is overseen by the Universal power-line association (UPA). More information about DHS can be found at www.upaplc.org.

## HD-PLC

High definition power line communication is communication system developed by Panasonic that uses high frequency signals over a power line to transmit data and digital media signals. The HD-PLC system transmits high-speed data signals using frequencies between 4 MHz to 28 MHz. The HD-PLC system supports the use of 128 bit encryption to ensure data privacy. It has a maximum transmission distance of approximately 150 meters.

## TVnet

TVnet is a smart coaxial based media distribution developed by Coaxsys. TVnet enables users to distribute many kinds of media throughout the home on a coaxial distribution system.

The TVnet multimedia distribution system operates by transferring media on unused frequency bands on the coaxial cable system. For example, the TVnet/C network uses the coax spectrum above 1 GHz. This ensures that TVnet signals will not interfere with channels received from another television service provider.

Because multiple devices may be networked on a TVnet system, the master control device in the TVnet system directs the flow of traffic on the network, thus enabling target devices to send and receive IP traffic.

The TVnet master sends control commands between target devices, which allows for the establishment and control of communication sessions. Essentially, the master serves as the arbiter on the coax, deciding which target TVnet device is permitted to communicate at which time. This prevents a problem on the network, ensuring that devices do not try to communicate over the top of one another.

The TVnet system uses an Ethernet type access structure that allows IP packets to be transferred between devices. The TVnet system uses RF modulation technology to allow IP packets (whether video, voice, or data) to be transmitted over coax. TVnet devices convert the digital signal into an analog signal, transmit the new analog signal over coax, and convert the signal back to digital once it reaches its destination.

Each device that will be part of the TVnet network requires a TVnet "target" adapter. This target will be located next to the Ethernet device, such as an IP set-top box, PC, or IP telephone. If the customer desires, they can also use the device's pass-through port to allow cable television signals to pass directly through the TVnet target adapter.

Figure 7.14 shows how TVnet systems can allow a coaxial distribution system to dynamically interconnect devices by connecting devices to coax and using unutilized coax spectrum for the purposes of IP networking. This example shows that a master coordinates the transmission of signals without interfering with the transmission between other devices. In this diagram, an IPTV head end streams a movie through a broadband modem to a television in another room. At the same time, a computer is connected

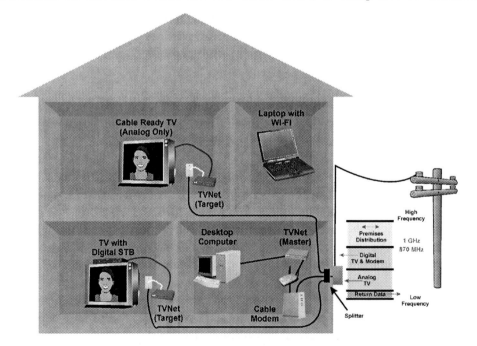

*Figure 7.14, TVNet Distribution System*

through a coax cable to the broadband modem to the Internet and a television is connected to a media server through the broadband modem. This example also shows that the master device in this coax system has the capability to setup connections and prioritize transmissions based on the type of media (such as real time video over web browsing data).

The TVnet/C is setup to allow eight target connections for each master unit, providing a total of 9 devices with the basic system. However, the TVnet system can be expanded to virtually an unlimited number of connections.

The user can also connect to multiple digital devices at any one connection point, so the number of devices that can be connected is far higher than nine. In practice, operators deploying in single-family homes typically deliver IPTV to 2-4 televisions. However, there are some applications such as IP video on demand or IPTV installations in multiple dwelling units (MDUs) that use far more Targets per Master than the typical 2-4.

There are two key types of equipment that are needed for a TVnet system; a TVnet Master and TVnet Targets. Alternatively, TVnet technology can be integrated into the modem or set-top box. No other software, switch settings or configuration setups are required.

The TVnet system typically achieves actual throughput rates of 70 to 100 Mbps, and over 200 Mbps has been demonstrated. An important consideration about data transmission rates is the difference between physical data transmission rates (the "PHY" rate) and actual data throughput. The throughput is affected by the amount of overhead (control) data and the amount of information that is lost via interference and the access control process. Because coaxial cable systems can transfer signals with minimal signal leakage, they are relatively immune from interfering signals and can have higher throughputs that other types of premises distribution networks.

## Multimedia over Coax Alliance (MoCA)™

Multimedia over Coax Alliance is a non-profit association that works to help develop and promote unified information about networking technologies, products and services that are primarily distributed over coax cabling systems within a building or premises. MoCA has developed a system for transferring multimedia signals over coax networks.

The MoCA system operates by transferring media on frequency bands between 860 MHz and 1.5 GHz using on the coaxial cable system [2]. The MoCA system uses 50 MHz transmission channel that transmit at 270 MHz. Each 50 MHz channel is subdivided into smaller channels using orthogonal frequency division multiplexing (OFDM) and the net throughput of each MoCA channel is 135 Mbps. Because the MoCA system can have multiple devices and it may co-exist with other systems, it can use up to 16 different channels that can be dynamically assigned and controlled. These channels can operate independently allowing MoCA systems to co-exist together.

The MoCA system is designed to prioritize the transmission of media to ensure a reliable quality of service (QoS) for various types of media. The MoCA system can schedule (reserve) bandwidth so it can provide different QoS levels. The MOCA system uses DES for link level encryption to ensure privacy between neighboring devices and systems.

The MoCA system was designed to operate with most existing home coaxial networks without requiring changes. Home coaxial networks commonly are setup as a tree structure and there may be several signal splitters used in the tree. To operate without any changes, signals from one device in the MoCA system must travel across splitters ("splitter jumping"). Because splitter jumping the transmits of a signal from a device connected to an output port to another device that is connected to a different output port of the coupler, the insertion loss (attenuation) is high.

The MoCA system enables splitter jumping by having a large link budget. Link budget is the maximum amount of signal losses that may occur between a transmitter and receiver to achieve an adequate signal quality level. The link budget includes splitter losses, cable losses and signal fade margins.

A potential challenge for bi-directional coaxial distribution systems is the use of coaxial amplifiers. Coaxial amplifiers are devices or assemblies that are used to amplify signals that are transmitted on coaxial cables. Coaxial amplifiers may offer unidirectional or bidirectional amplification capabilities. The frequency range and coaxial amplifiers is typically chosen to match the signals that are transmitted on the coaxial cable.

Coaxial amplifiers may be used on home television systems that have long cable connections or that have many television ports (many splitters). There are two types of coaxial amplifiers; drop amplifiers and inline amplifiers. Drop amplifiers are signal amplifiers that have been inserted after a communication drop to a building and before other devices that are connected to the premises network. Inline amplifiers are signal amplifiers that have been inserted into a premises network where it is located between (inline with) devices. Inline amplifiers can be a potential block for coaxial distribution signals. In the United States, about 2% of homes have in line amplifiers and approximately ½ of these homes may have challenges using a PDN system [3].

Figure 7.15 shows how a MoCA systemizes a coaxial distribution system to dynamically interconnect devices by connecting devices to coax and using frequencies above 860 MHz. This example shows that the MoCA system is designed to operate in an existing cable television system by allowing signals to jump across and through existing splitters.

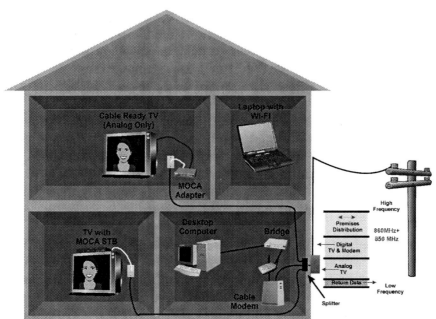

*Figure 7.15, MOCA Distribution System*

## 802.11 Wireless LAN

802.11 Wireless LAN (commonly called "Wi-Fi") is a set of IEEE standards that specify the protocols and parameters that are used for commonly available wireless LANs that use radio frequency for transmission. The use of Wi-Fi systems for IPTV distribution offers the advantage of requiring "no new wires." Wireless LAN systems operate in the unlicensed frequency bands.

Until recently, reliable distribution of IPTV signals over consumer-grade Wi-Fi systems have not been possible. Wi-Fi is fundamentally an unreliable medium due to that fact that it is shared among multiple users over the unlicensed RF spectrum. The key challenges that face providers and subscribers running IPTV over Wi-Fi is radio signal interference from other sources, signal quality levels, range and data transmission rates.

The use of unlicensed frequency bands can result in interference from other devices that used the same frequency bands. This includes interference that may come from other Wi-Fi systems that operating nearby or microwave ovens or other obstructions that may dynamically appear and cause problem in the transmission of video over Wi-Fi.

Figure 7.16 shows typical types of unlicensed radio transmission systems that can cause interference with WLAN systems. This example shows that there are several different communication sessions that are simultaneously operating in the same frequency band and that the transmission of these devices are not controlled by any single operator. These devices do cause some interference with each other and the types of interference can be continuous, short-term intermittent or even short bursts. For the video camera (such as a wireless video baby monitor), the transmission is continuous. For the cordless telephone, the transmission occurs over several minutes at a time. For the microwave oven, the radio signals (undesired) occur for very short bursts only when the microwave is operating. For the wireless head-

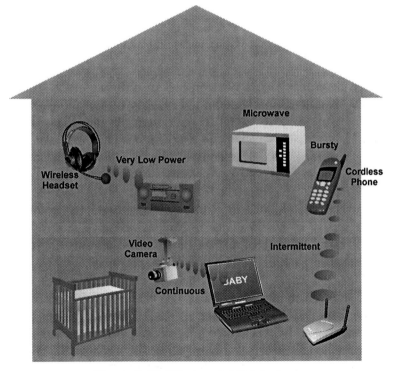

*Figure 7.16, Wireless LAN Interference*

set, the transmission occurs for relatively long periods of time but the power is very low so interference only occurs when the wireless headset gets close to WLAN devices.

Radio interference can cause lost packets. This results in sub-standard video that freezes or has an unacceptable amount of artifacts. Sources of WLAN interference are varied and many including microwave ovens, baby monitors, cordless telephones, neighboring Wi-Fi networks and even people.

In addition to overcome potential interference issues, Wi-Fi systems that are used for IPTV systems, they must be able to support quality of service for different types of media. If it cannot, a user watching an IPTV stream might have their viewing interrupted when another user is downloading a file over the same Wi-Fi network. Wi-Fi systems that are used for IPTV multimedia home network may use a combination of 802.11e quality of service, 802.11n MIMO and smart antenna systems in increase the reliability and performance of the Wi-Fi system.

## *802.11e Quality of Service*

The 802.11e wireless local area network is an enhancement to the 802.11 series of WLAN specifications that added quality of service (QoS) capabilities to WLAN systems. The 802.11e specification modifies the medium access control (MAC) layer to allow the tracking and assignment of different channel coding methods and flow control capabilities to support different types of applications such as voice, video, and data communication.

Medium access control (MAC) is the processes used by communication devices to gain access to a shared communications medium or channel. A MAC protocol is used to control access to a shared communications media (transmission medium) which attaches multiple devices.

MAC is part of the OSI model Data-Link Layer. Each networking technology, for example Ethernet, Token Ring or FDDI, have drastically different protocols which are used by devices to gain access to the network, while still providing an interface that upper layer protocols, such as TCP/IP may use without regard for the details of the MAC protocol that is used. In short, the MAC provides an abstract service layer that allows network layer protocols to be indifferent to the underlying details of how network transmission and reception operate.

The 802.11e wireless local area network is an enhancement to the 802.11 series of WLAN specifications that added quality of service (QoS) capabilities to WLAN systems. The 802.11e specification modifies the medium access control (MAC) layer to allow the tracking and assignment of different channel coding methods and flow control capabilities to support different types of applications such as voice, video, and data communication.

Enhanced distributed channel access is a medium access control (MAC) system that is used in the 802.11e WLAN system that enables the assignment of priority levels to different types of devices or their applications. The prioritization is enabled through the assignment of different amount of channel access back-off times where the amount of back-off time depends on the packet priority. EDCA defines four priority levels (four access categories) for different types of packets.

To coordinate and prioritize packets between devices within an 802.11 network, a hybrid coordination function controlled channel access (HCCA) can be used. The HCCA uses a central arbiter (access coordinator) that assigns the access categories for different types of packets. The central arbiter receives transmission requests from all the devices communicating in the 802.11 system so it can assign and coordinates transmission and tasks that are assigned to other devices. The HCCA process can guarantee reserved bandwidth for packets classified by using the EDCA process.

## 802.11n Multiple Input Multiple Output (MIMO)

The 802.11n wireless local area network is an enhancement to the 802.11 series of WLAN specifications that adds multiple input and multiple output (MIMO) capability to WLAN systems. The 802.11n specification modifies the medium access control (MAC) layer to allow channel bonding (channel combining) and this increases the available data transmission rate while increasing the reliability (robustness) of the Wi-Fi system..

Multiple input multiple output is the combining or use of two or more radio or telecom transport channels for a communication channel. The ability to use and combine alternate transport links provides for higher data transmission rates (inverse multiplexing) and increased reliability (interference control).

Figure 7.17 shows how a multiple input multiple output (MIMO) transmission system transmits signals over multiple paths to a receiver where they are combined to produce a higher quality signal. This example shows that even if an object or signal distortion occurs in one of the transmitted paths, data can still be transmitted on alternative paths.

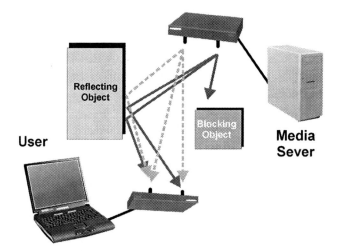

Figure 7.17, Multiple Input Multiple Output (MIMO)

## Smart Antenna System

A smart antenna system uses active transmission components to allow the forming or selection of specific antenna patterns. Smart antennas may have multibeam capability that allows for the reuse of the same frequency in the same radio coverage area. Using smart antenna systems, the transmission energy of a signal can be directed in a specific direction that does not result in the interference to other signals that are operating in the same general area.

Wi-Fi systems that have smart antenna capability can alter multicast traffic by directing the traffic to a specific receiver, forcing that receiver to provide an acknowledgement. This way the system knows if the video transmission was received and the quality of the link.

Figure 7.18 shows how WLAN systems can be improved through the use of directional transmission and media prioritization to provide improved performance. This example shows that a wireless access point has been enhanced to allow the transmission of signals using directional antennas so that signals can be sent to specific devices. In this diagram, one of the best paths between the access point (AP) and Wi-Fi device (e.g. Wi-Fi Television) is not direct as a metal art object that is located between the AP and the Wi-Fi device reflects the radio signal. This example also shows that this WLAN system has the capability to prioritize transmission based on the type of media (such as real time video over web browsing data).

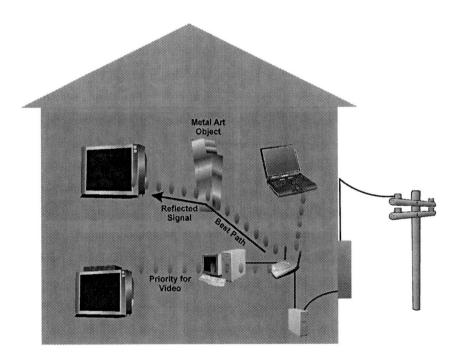

*Figure 7.18, Smart Wireless LAN*

References:

[1]. "MoCA," PDN requirements.

[2]. "IPTV Crash Course," Joseph Weber, Ph.D. and Tom Newberry, Mc-Graw-Hill, New York, 2007, pg. 172.

[3]. "MOCA" Inline amplifiers.

# Chapter 8

## IPTV End User Devices

IPTV end user devices adapt an IP communications medium to a format that is accessible by the end user. IP end user devices are commonly located in a customer's home to allow the reception of video signals on a television. The key functions for end user devices include interfacing to the communication network, selecting and decoding channels, processing the media into a form usable by humans and providing controls that allow the user to interact with the device.

The network interface for IPTV end user devices allow it to receive IP data. These interfaces include Ethernet, DSL, cable, optical or wireless connections. The device must separate out (demultiplex) the communication channel from the network connection (if necessary) and separate out the component parts of the television signal (video, audio and data). The underlying media is then decoded and decrypted (unscrambled). The media is then converted (rendered) into a form that can be displayed or heard by the end user. A program guide and menu system is provided to allow the user to navigate and select features and services.

Figure 8.1 shows the basic functions of an IPTV end user viewing device. This diagram shows that the end user device has a network interface, signal processing, decoding, rendering and user interface selections. The network interface may receive and transmit signals on broadband CATV, satellite, DSL modem, wireless or optical systems. The signal processing section receives, selects and demultiplexes the incoming channels. After the channels are received, the channel may require decoding (decryption) for scrambled channels. The STB then converts the data into signals that can be displayed to the viewer (rendering). The STB has a user interface that allows the system to present information to the user (such as the program guide) and to allow the user to interact (select channels) with the STB.

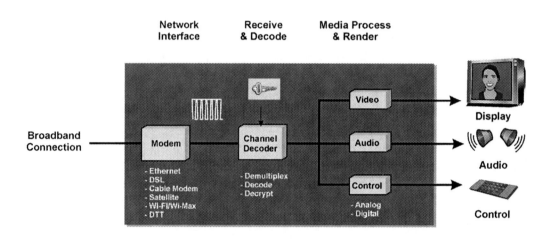

Figure 8.1, IPTV User Device Functions

The key types of end user devices used in IPTV systems include set top boxes (STBs), IP televisions, multimedia mobile telephones, multimedia computers and portable media players.

## Set Top Boxes (STB)

A set top box is an electronic device that adapts a communications medium to a format that is accessible by the end user. Set top boxes are commonly located in a customer's home to allow the reception of video signals on a television or computer.

## Internet Protocol Set Top Box (IP STB)

IP set top boxes are devices that convert IP packet signals into analog or digital television formats (e.g. NTSC, PAL, HDMI). Using IP STBs, it is possible to use standard televisions for viewing television channels that are sent over a data network such as DSL or over the Internet.

An IP STB is basically a dedicated mini computer, which contains the necessary software and hardware to convert and control IP television signals. IP STBs must convert digital broadband media channels into the television (audio and video signals) and decode and create the necessary control signals that pass between the IP STB and system headend equipment.

## Download and Play (DP STB)

Download and play set top boxes are devices that can select, completely transfer (download) and play media programs. Because download and play STBs wait until programs are completely downloaded before playing, they can transfer programs through unmanaged networks (such as the Internet) and play the programs without errors or delays.

## Hybrid STB

A hybrid set top box is an electronic device that adapts multiple types of communications mediums to a format that is accessible by the end user. Hybrid set top boxes are commonly located in a customer's home to allow the reception of video signals on a television or computer. The use of HSTBs allows a viewer to get direct access to broadcast content from terrestrial or satellite systems in addition to accessing other types of systems such as interactive IPTV via a broadband network.

## IP Television (IP Television)

IP televisions are television display devices that are specifically designed to receive and decode television channels through IP networks without the need for adapter boxes or media gateways. IP televisions contain embedded software that allows them to initiate and receive television signals through that network using multimedia session protocols such as SIP.

An IP television has a data connection instead of a television tuner. IP televisions also include the necessary software and hardware to convert and control IP television signals into a format that can be displayed on the IP television (e.g. picture tube or plasma display).

## Mobile Video Telephones

Mobile video telephones have multimedia capabilities that allow them to receive and display video signals. Mobile telephones usually have limited information processing capability, small displays and may have restricted access to Internet services.

Multimedia mobile telephones contain embedded software that allows them to initiate and receive multimedia communication sessions through the Internet. Because of the limited bandwidth and higher cost of bandwidth for mobile telephones, media players in mobile telephones may use compression and protocols that are more efficient than what are used by standard IP television systems. To increase the efficiency, mobile telephone data sessions may be connected through gateways that compress media signals and convert standard control protocols to more efficient and robust control protocols. This may cause some incompatibilities or unexpected operation for mobile television systems.

## Multimedia Computers

A multimedia computer is a data processing device that is capable of using and processing multiple forms of media such as audio, data and video.

Because many computers are already multimedia and Internet ready, it is often possible to use a multimedia computer to watch IP television through the addition or use of media player software. The media player must be able to find and connect to IP television media servers, process compressed media signals, maintain a connection, and process television control features.

Control of IP television on a multimedia computer can be performed by the keyboard, mouse, or external telephone accessory device (such as a remote control) that can be connected to the computer through an adapter (such as an infrared receiver). The media player software controls the sound card, accessories (such as a handset), and manages the call connection.

To securely present IPTV signals on a multimedia computer (to ensure the television program is not stored, copied and distributed), a soft client is used. A soft client is a software program that operates on a computing device (such as a personal computer) that can request and receive services from a network for specific applications. The soft client coordinates the transmission, storage, decoding (if the signal is encrypted) and displaying of the program to the monitor.

### Portable Media Players

A portable media player is a self-contained device and associated software application that can convert media such as video, audio or images into a form that can be experienced by humans. Portable media players may have small display areas with limited resolution capability, reduced navigation controls and limited signal processing and decoding capabilities.

## IP STB Capabilities

IP STB capabilities include network connection types, display capability, media processing (video, audio and graphics), security, communication protocols, software applications, accessories, middleware compatibility, media distribution and upgradeability.

## Network Connection Types

Network connection types are the physical and system connections that may be used to provide communication services. Set top boxes often have multiple network connection types and there may be multiple service providers who can provide services through the STB. Common network connection types for IPTV viewing devices include Ethernet, DSL, Cable, Satellite and DTT.

## Display Capability

Display capability is the ability of a device to render images into a display area. Display capabilities for IP STB include size and resolution (SD or HD), the type of video (interlaced or progressive) and display positioning (scaling and displaying multiple sources). Display capabilities for television systems are characterized in the MPEG industry standards and the sets of capabilities (size and resolution) are defined as MPEG profiles.

## Video Scaling

Video scaling is the process of adjusting a given set of video image attributes (such as pixel location) so they can be used in another format (such as a smaller display area). Video scaling can be used to reduce the size of a display to an area that allows for placing it or surrounding it within a video picture or within in a graphics image. When a video program is placed within a graphic image, it is called picture in graphics. Picture in graphics may be used to display other media items (such as a channel menu).

## Dual Display (Picture in Picture)

Picture in picture is a television feature that allows for the display of one or more video images within the display area of another video display. Picture in picture typically allows a viewer to watch two (or more) programs simultaneously. To obtain two picture sources, it usually requires that the STB has multiple receivers.

## Security

Security for set top boxes is the ability to maintain its normal operation without damage and to ensure content that is accessed by the STB is not copied or used in an unauthorized way by the user. Set top boxes commonly include smart cards and security software to ensure the content is used in its authorized form.

## Smart Card

A smart card is a portable credit card size device that can store and process information that is unique to the owner or manager of the smart card. When the card is inserted into a smart card socket, electrical pads on the card connect it to transfer information between the electronic device and the card. Smart cards are used with devices such as mobile phones, television set top boxes or bank card machines. Smart cards can be used to identify and validate the user or a service. They can also be used as storage devices to hold media such as messages and pictures.

Smart card software can be embedded (included) in the set top box to form a virtual smart card. A virtual smart card is a software program and associated secret information on a users device (such as a TV set top box) that can store and process information from another device (a host) to provide access control and decrypt/encrypt information that is sent to and/or from the device.

## DRM Client

A digital rights management client is a computer, hardware device or software program that is configured to request DRM services from a network.

An example of a DRM client is a software program (module) that is installed (loaded) into a converter box (e.g. set top box) that can request and validate information between the system and the device in which the software is installed.

## Secure Microprocessor

A secure microprocessor is a processing device (such as an integrated circuit) that contains the processes that are necessary to encrypt and decrypt media. Secure microprocessors contain the cryptographic algorithms such as DES, AES or PKI. The secure microprocessor can be a separate device or it can be a processing module that is located within another computing device (such as a DSP).

## Media Processing

Media processing is the operations used to transfer, store or manipulate media (voice, data or video). The processing of media ranges from the playback of voice messages to modifying video images to wrap around graphic objects (video warping). Media processing in set top boxes includes video processing, audio processing and graphics processing.

Video processing is the methods that are used to convert and/or modify video signals from one format into another form using signal processing. An example of video processing the decoding of MPEG video and the conversion the video into a format that can be displayed on a television monitor (e.g. PAL or NTSC video).

Audio processing is the methods that are used to convert and/or modify audio signals from one format into another form using signal processing. An example of audio processing is the decoding of compressed audio (MP3 or AAC) and conversion into multiple channels of surround sound audio (5.1 audio).

Graphics processing is the methods that are used to convert and/or modify image objects from one format into another form. An example of graphics processing the conversion of text (e.g. subtitles) into bitmapped images that can be presented onto a television display (on screen display).

## Communication Protocols

Communication protocols are sets of rules and processes that manage how transmissions are initiated and maintained between communication points. Communication protocols define the format, timing, sequence, and error checking used on a network or computing system. Communications protocols vary in complexity ranging from simple file transfer protocols to reservation protocols that setup and manage the reliability of connections through a network. There are many types of protocols (with multiple versions and options) used by IP set top boxes. Some of the common protocols used in IP STB include; TCP/IP, DHCP, SNMP, MMS, RTSP, TS, IGMP, PPPoE, TFTP and HTTP.

Transmission control protocol and Internet protocol (TCP/IP) coordinates the transmission and reception of packets through IP networks. Dynamic host configuration protocol (DHCP) obtains the IP address that is used during communication between the STB and the system. Simple network management protocol (SNMP) is used to capture and change equipment settings. Microsoft media server protocol (MMS) coordinates the transfer of multimedia objects and files. Real time streaming protocol (RTSP) provides for media flow control to allow users to select, play, stop and move through media programs. Transport stream (TS) protocol coordinates how multiple media channels (video, audio and data) share the same transmission channel. Internet group management protocol (IGMP) allows multiple users to share the same stream of media (e.g. watch a TV program from a nearby router instead of setting a direct connection for each user). Point-to-point protocol over Ethernet (PPPoE) is used to setup and coordinate a communication link over an Ethernet connection. Trivial file transfer protocol (TFTP) is used to transfer programs and configuration files between the system and the STB. Hypertext transfer protocol (HTTP) is used to select and coordinate transfer media and data files.

## Software Applications

A software application is a software program that performs specific operations to enable a user to apply the software to their specific needs or problems. Software applications in set top boxes may be in the form of embedded applications, downloaded applications or virtual applications.

Embedded applications are programs that are stored (encapsulated) within a device. An example of an embedded application is a navigation browser that is included as part of a television set top box. Downloaded applications are software programs that are requested and transferred from the system to the user device when needed. A loader application (the loader is an embedded application) is used to request and transfer applications from the system. Virtual applications are software instructions that are written in another language to perform applications using an interpreter program (e.g. Javascript).

## Accessories

IPTV accessories are devices or software programs that are used with IPTV systems or services. Examples of IPTV accessories include remote controls, gaming controllers and other human interface devices that are specifically designed to be used with IPTV systems and services. These accessories may have dedicated connection points (such as game controllers) or they may share a standard connection (such as a USB connection).

## Middleware Compatibility

Middleware compatibility is the ability of a device to accept software programs (clients) that interface the device to the system host (servers). A middleware client is a software module that is installed in a device that is configured to request and deliver media or services from a server (e.g. to request television programs from media network).

## Upgradability

Upgradability is the ability of a device or system to be modified, changed or use newer components and/or as technology innovations become available. The ability to upgrade the capabilities of a set top box may be performed by software downloads or through the use of software plug-ins.

## Plug-In

A plug-in is a software program that works with another software application to enhance its capabilities. An example of a plug-in is a media player for a web browser application. The media player decodes and reformats the incoming media so it can be displayed on the web browser.

## Media Portability

Media portability is the ability to transfer media from one device to another. Media portability can range from stored media locally in a hard disk (for personal video recorder) or by shared media through home connections (such as a premises distribution network).

# End User Device Operation

An IP set top box (IP STB) is an electronic device that adapts a broadband communications signals into a format that can be displayed on a television. The basic functions of an Internet IPTV set top box include selecting and connecting to a broadband connection, separating, decoding and processing media and interfacing the media and controls with the user.

Figure 8.2 shows a functional block diagram of an IP STB. This diagram shows that an IP STB typically receives IP packets that are encapsulated in Ethernet packets. The IP STB extracts the IP packets to obtain the transport stream (TS). The channel decoder detects and corrects errors and provides the transport stream to the descrambler assembly. The descrambler assembly receives key information from either a smart card or from an external conditional access system (e.g. via a return channel). Using the key(s), the STB can decode the transport stream and the program selector can extract the specific program stream that the user has selected. The IP STB then demultiplexes the transport stream to obtain the program information. The program table allows the IP STB to know which streams are for video, audio and other media for that program. The program stream is then divided into its elementary streams (voice, audio and control) which is supplied to a compositor that create the video signal that the television can display.

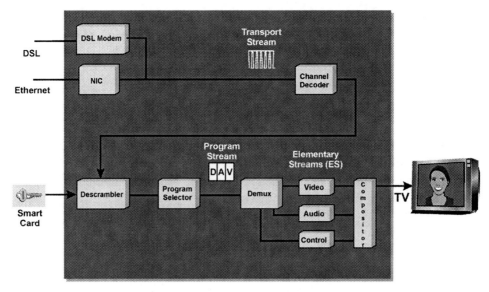

Figure 8.2, IPTV STB

## Network Interface

A network interface is the point of connection between equipment and a communication network. A network interface is the physical and electrical boundary between two separate communication systems. The network interface on set top box may require network selection (if there are multiple networks) and channel selection (if there are multiple channels on a network connection).

## Network Selection

Network selection is the process of identifying and connecting to a network line or system. IP STB may have one or more network connection options that include Ethernet, DSL modem, cable, Wi-Fi, satellite, DTT or other network system. The choice of network connection may include availability of a network signal (is it attached to the network), preferred connection type (higher speed connections) and cost considerations.

## Channel Selection

Channel selection is the process of identifying, tuning or adapting and receiving a communication signal. Some networks, such as a cable television network, have multiple channels available on a communication line. For simulcast systems (such as CATV systems), the STB uses a tuner to receive a specific frequency channel.

## Return Path

A return path is the medium that is used by information to return to the origin point of the communication signal. IP set top boxes use a return path to permit users to select programs and feature options. The return path may be on the same broadband network that provides IPTV service (e.g. DSL or cable modem) or it may be sent via another path such as a dial-up telephone line.

## Demultiplexing

Demultiplexing is a process that separates individual channels from a common transport or transmission channel. STBs perform demultiplexing to separate out logical channels (such as a virtual data channel) and media channels (video, audio and data).

## Decoding

Decoding is the process of converting encoded media or data words into its original signal. Decoding involves extracting control signals (e.g. link management) and decrypting the data (if the data has been encrypted).

## Rendering

Rendering is the process of converting media into a form that a human can view, hear or sense. An example of rendering is the conversion of a data file into an image that is displayed on a computer monitor. Rendering is performed by decoding the compressed program media (e.g. MPEG) and encoding it into a form that can used by a display (TV) or audio device (speakers).

## Video Coding

Video coders convert video media streams or files back into their original or another form. Once the received video is decompressed, it is converted (coded) into a form that can be displayed to the user such as analog video (NTSC or PAL) or digital video (HDMI/DVI).

## Audio Coding

Audio coders convert audio media streams or files back into their original (or near original) form. Once the received audio is decompressed, it is converted (coded) into a form that can be played to the user such as analog audio or digital audio (S/PDIF).

## Graphic Processing

Graphics processing is the methods that are used to convert and/or modify image objects from one format into another form. An example of graphics processing the conversion of text into bitmapped images that will be presented onto a display.

IPTV systems use graphics processing for navigation menus, alpha blending, and other features such as closed captioning. Alpha blending is the combining of a translucent foreground color with a background color to produce a new blended color. Closed captioning is a service that adds additional information (e.g. text captions) to television programs. Closed captioning information is typically provided by an additional signal that requires a closed captioning decoder to receive and display the captions on the television screen.

## User Interface

A user interface is a portion of equipment or operating system that allows the equipment to interact with the user. The user interface includes visual, audio and interface options such as menu screens, program guides, remote controls, control transmitters (IR blasters), web interface, game port and USB accessory ports.

## On Screen Display (OSD)

On screen display is the insertion of graphics or images onto the display portion of a screen. The graphics insertion typically occurs at the graphics card or set top box assembly that creates the signals for the display assembly (such as a computer monitor or a television set).

The graphics presented by the set top box is created as a display layer that is merged with other display layers such as the background layer and video layer. The graphics display layer may completely overlay the underlying video and background layers or it may be made to be translucent by alpha blending. The use of alpha blending allows the viewer to continue to see the programming through the graphic while they view the graphic and make their selections.

## Electronic Programming Guide (EPG)

Electronic programming guides are an interface (portal) that allows a customer to preview and select from possible lists of available content media. EPGs can vary from simple program selection guides to interactive filters that dynamically allow the user to search through program guides by theme, time period, or other criteria.

## Remote Control

A remote control is a device or system that can be used to control services or a device from a distance. For IPTV set top boxes, remote controls create commands that are used to initiate trick play services. Trick play is the ability of a system to provide for remote control operations during the streaming of a media signal. Trick mode may create a low-resolution display when users scan through a segment of media (such as fast sliding or slow looking).

## Wireless Keyboard

A wireless keyboard is a physical device that allows a user to enter data to a computer or other electronic device that is transferred to another device (such as a computer or set top box) via a wireless connection (e.g. radio or infrared).

## Infrared Blaster

An infrared blaster is a device that can send (transmit) information on an Infrared carrier signal. IR blasters may be used to by set top boxes to control other accessories (such as tape or DVD players).

## Web Interface

A web interface is the software program that enables a device, system or service to be controlled via a web page or Internet access device. A set top box may have the capability to allow a user to connect to it via a web page to setup operations such as personal video recording times.

## Game Port

A game port is a connection that allows for the attaching of joysticks and other interactive devices. Older game ports used a relatively large DB15 connector and new game devices use a standard USB port.

## Universal Serial Bus (USB)

Universal serial bus is an industry standard data communication interface. The USB was designed to replace the older UART data communications port. There are two standards for USB. Version 1.1 that permits data transmission speeds up to 12 Mbps and up to 127 devices can share a single USB port. In 2001, USB version 2.0 was released that increases the data transmission rate to 480 Mbps.

# Premises Distribution

A premises distribution network is the equipment and software that is used to transfer data and other media in a customer's facility or home. A PDN is used to connect terminals (computers) and media devices (such as TV set top boxes) to each other and to wide area network connections. PDN systems may use wired Ethernet, wireless LAN, powerline, coaxial and phone lines to transfer data or media.

# Interfaces

An interface is a device or circuit that is used to interconnect two pieces of electronic equipment. Interfaces are a boundary or transmission point that is common betweento two or more similar or dissimilar command and control systems.

## Network Connections

A network connection is the point of connection between equipment and a communication network. A network connection is the physical and electrical boundary between two separate communication systems.

## Ethernet

Ethernet is a packet based transmission protocol that is primarily used in LANs. Ethernet is the common name for the IEEE 802.3 industry specification and it is often characterized by its data transmission rate and type of transmission medium (e.g., twisted pair is T and fiber is F). Ethernet can be provided on twisted pair, coaxial cable, wireless, or fiber cable. Ethernet is a common data connection output connection offered by broadband modems (DSL or cable modems).

## Digital Subscriber Line (DSL)

Digital subscriber line is the transmission of digital information, usually on a copper wire pair. Although the transmitted information is in digital form, the transmission medium is an analog carrier signal (or the combination of many analog carrier signals) that is modulated by the digital information signal. DSL data transmission rates range from approximately 1 Mbps to 52 Mbps per line and multiple DSL lines can be combined (bonded) to provide higher combined data rates (100 Mbps+).

## Satellite Connection (Optional)

A satellite connection is a coaxial line that carries satellite RF channels. The frequency of satellite connections is typically above the television frequency band. If a satellite connection if available in the STB, the STB must contain a satellite receiver that can select which satellite RF signal to demodulate and which program channels to demultiplex from the RF carrier channel.

## DTT Connection (Optional)

Digital terrestrial television is the broadcasting of digital television signals using surface based (terrestrial) antennas. DTT is also called digital video broadcasting terrestrial (DVB-T).

## Cable TV Connection (Optional)

A cable TV connection is a coaxial line that carries RF channels between the set top box and a cable TV system. A cable TV connection requires a tuner to select the RF channel and a channel demultiplexer to separate out the video and audio channels from the RF carrier.

## TV RF Output

A TV radio frequency output is a connection point on a television device (such as a STB or DVD player) that provides television signals on an RF channel (such as channel 3 or 4).

## Video

Video outputs can be analog or digital form and they can be in component (e.g. color component) form or combined (composite form). A set top box also may contain video inputs can be used to connect webcams for security monitoring or legacy devices such as a VHS tape player

## Composite Video

Composite video is a single signal that contains all the information necessary to create a video display. Composite video signals are commonly in NTSC or PAL form.

## Component Video

Component video consists of separate signals to represent color. Component video may be in various forms including red, green, and blue (RGB) or intensity and color difference signals (YPbPr). The RGB is commonly used for computing devices and the YPbPr are used in television systems.

## S-Video

S-video is a set of electrical signals that represent video luminance and color information. The separation of intensity and color information provides for a higher quality video signal than composite video.

Figure 8.3 shows the pin configurations and signal types used on the S-video connectors. This example shows that the small connector (mini 4 pin) contains a C (chrominance color) video pin #4 and a Y (luminance intensity) video pin #3 and the other two pins are used for a ground (return path) connection. The standard 7 pin S-Video connector contains a C (chrominance color) video pin #4 and a Y (luminance intensity) video pin #3 along with a tuner control data line pin #6 and a tuner clock control line #5 and the remaining two pins (#1 and #2) are used for a ground (return path)

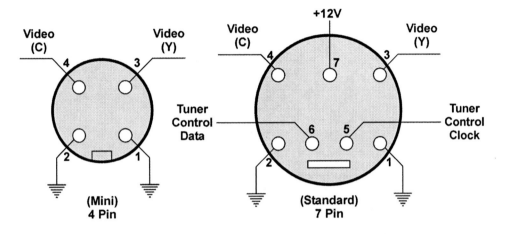

Figure 8.3, S-Video Connector

## SCART

A SCART connector is an industry standard interface connector that may be used in television devices to carry video and audio signals. The connector primarily is used to provide composite and component video connections between devices. The connector has 21 pins of which 20 pins are in the center and one pin is the shield of the connector. The connector is designed to allow for cascading (daisy chaining) multiple connectors to each other. The SCART connector is also known as the Peritel connector and Euroconnector and the connector is required on consumer devices in some countries (such as France).

The SCART connector does include data lines and switch control lines that allows for the remote powering on of the television and other devices. It also includes a control line that allows for the sensing and control of widescreen mode.

Figure 8.4 shows a diagram of a SCART connector and some of its pin functions. This diagram shows that the SCART connector has 21 pins – 20 in the center block and the shield of the connector is a common ground. This diagram shows that the SCART connector can provide component, composite and S-video signals.

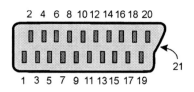

| Pin | Function | Pin | Function |
|---|---|---|---|
| 1 | Audio Out (right) | 12 | D2B Output |
| 2 | Audio In (right) | 13 | Red Ground |
| 3 | Audio Out (left) | 14 | D2B Ground |
| 4 | Audio Ground | 15 | Red |
| 5 | Blue Ground | 16 | Fast Switch |
| 6 | Audio In (left) | 17 | Composite Video Out Ground |
| 7 | Blue | 18 | Composite Video In Ground |
| 8 | Widescreen Select | 19 | Composite Video Out |
| 9 | Green Ground | 20 | Composite Video In |
| 10 | D2B Input | 21 | Common Ground |
| 11 | Green | | |

Figure 8.4, SCART Connector Diagram

## IEEE 1394

IEEE 1394 is a personal area network (PAN) data specification that allows for high-speed data transmission (400 Mbps). The specification allows for up to 63 nodes per bus and up to 1023 busses. The system can be setup in a tree structure, daisy chain structure or combinations of the two. IEEE 1394 also is known as Firewire or I.Link.

# High Definition Multimedia Interface (HDMI)

HDMI is a specification for the transmission and control of uncompressed digital data and video for computers and high-speed data transmission systems. HDMI is a combination of a digital video interface (DVI) connection along with the security protocol HDCP. HDMI has the ability to determine if security processes are available on the DVI connection and if not, the HDMI interface will reduce the quality (lower resolution) of the digital video signal. The HDMI connector provides for very high speed data transmission at 10.2 Gbps using transition minimized differentiated signaling (TMDS®) balanced lines.

Figure 8.5 shows an HDMI connector diagram. This diagram shows that the HDMI connector is has 19 pins that are used to carry digital signals and is 13.9 mm wide by 4.45 mm high. The HDMI connector has 3 high-speed TDMS® data connections (0, 1 and 2).

| Pin | Function | Pin | Function |
|-----|----------|-----|----------|
| 1 | TMDS Data 2+ | 11 | TMDS Clock Shield |
| 2 | TMDS Data 2 Shield | 12 | TMDS Clock- |
| 3 | TMDS Data 2- | 13 | CEC |
| 4 | TMDS Data 1+ | 14 | Reserved |
| 5 | TMDS Data 1 Shield | 15 | SCL |
| 6 | TMDS Data 1- | 16 | SDA |
| 7 | TMDS Data 0+ | 17 | DDC/CEC Ground |
| 8 | TMDS Data 0 Shield | 18 | +5V |
| 9 | TMDS Data 0- | 19 | Hot Plug Detect |
| 10 | TMDS Clock+ | | |

Figure 8.5, HDMI Connector Diagram

## Audio

Audio is a signal that is composed of frequencies that can be created and heard by humans. The frequency range for an audio signal typically ranges from 15 Hz to 20,000 Hz. Audio outputs can be in baseband (analog audio signal form) or in digital form.

## Baseband Audio

Baseband audio is the fundamental frequency components that contain the information of the audio signal. Baseband stereo audio is provided on many STBs with 5.1 channel surround sound options provided via the digital audio connections.

## Sony Philips Digital Interface (S/PDIF)

S/PDIF is a digital audio connection format that allows for the transmission of up to 6 channels of digital audio (5 audio and 1 sub-audio). All 6 channels time share a single serial data connection.

## Wireless

Wireless connections can be used to provide audio and video signals by radio signals. The common wireless types that are used in set top boxes include wireless LAN and ultra wideband (UWB). The use of wireless systems for IPTV distribution offers the advantage of requiring "no new wires."

## Wireless LAN (WLAN)

A wireless local area network (WLAN) was primary designed to allow computers and workstations to communicate with each other using radio signals. There are several versions of WLAN with data transmission rates ranging from 2 Mbps to 54 Mbps.

Until recently, reliable distribution of IPTV signals over consumer-grade Wi-Fi systems have not been possible. Wi-Fi is fundamentally an unreliable medium due to that fact that it is shared among multiple users over the unlicensed RF spectrum. The key challenges that face providers and subscribers running IPTV over Wi-Fi is radio signal interference from other sources, signal quality levels, range and data transmission rates.

The use of unlicensed frequency bands can result in interference from other devices that used the same frequency bands (such as microwave ovens and baby monitors). This includes interference that may come from other Wi-Fi systems that are operating nearby or microwave ovens or other obstructions that may dynamically appear and cause problem in the transmission of video over Wi-Fi.

In addition to overcome potential interference issues, Wi-Fi systems that are used for IPTV systems must be able to support quality of service for different types of media. If it cannot, a user watching an IPTV stream might have their viewing interrupted when another user is downloading a file over the same Wi-Fi network. Wi-Fi systems that are used for IPTV multimedia home network may use a combination of 802.11e quality of service, 802.11n MIMO and smart antenna systems in increase the reliability and performance of the Wi-Fi system.

## Ultra Wideband (UWB)

Ultra wideband is a method of transmission that transmits information over a much wider bandwidth (perhaps several GHz) than is required to transmit the information signal. Because the UWB signal energy is distributed over a very wide frequency range, the interference it causes to other signals operating within the UWB frequency band is extremely small. This may allow the simultaneous operation of UWB transmitters and other existing communication systems with almost undetectable interference.

Figure 8.6 shows a sample front panel of an IP set top box. This diagram shows that the front panel usually contains a visual display, keypad controls, infrared sensor and an accessory connector. This example shows that the visual display shows the selected channel and connection status. The keypad controls include power, menu selection and navigation keys. The infrared sensor is used to allow a remote control or a wireless keyboard to control the STB. The accessory connection is a USB connector that allows smart accessories to be connected to the set top box.

Figure 8.6, IP STB Front Panel View

Figure 8.7 shows a sample rear panel of an IP set top box which contains network connections, video connections, audio connections and a smart card socket, This example shows that the network connection options include Ethernet connection, DSL connection, DTT, Satellite and Cable TV. The video options include RF TV out, component video (red, green and blue), S-video and composite video. The audio outputs include left and right audio.

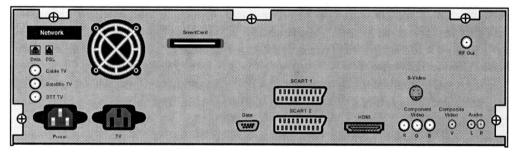

Figure 8.7, IP STB Real Panel

# Chapter 9

# Digital Rights Management (DRM)

Rights management is a process of organization, access control and assignment of authorized uses (rights) of content. Rights management may involve the control of physical access to information, identity validation (authentication), service authorization, media protection (encryption) and usage monitoring (enforcement). Rights management systems are typically incorporated or integrated with other systems such as content management system, billing systems, and royalty management.

Rights management systems are an implementation of the business and operations aspects of rights management. The rights managed by rights management systems can be affected by legal rights, transactional rights and implicit rights. Legal rights are actions that are authorized to be performed by individuals or companies that are specified by governments or agencies of governments. Transactional rights are actions or procedures that are authorized to be performed by individuals or companies that granted as the result of a transaction or event. An example of a transactional right is the authorization to read and use a book after it is purchased in a bookstore. Implicit rights are actions or procedures that are authorized to be performed based on the medium, format or type of use of media or a product.

Rights management systems are typically setup to protect intellectual property and to assist in the valuation and collection of fees for the sale of rights of the intellectual property. Intellectual property is intellect that has been converted into some form of value. Intellectual property may be represented in a variety of forms and the copying, transfer and use of the intellect may be protected or restricted.

Property in a rights management system requires an identification and description of property items. The owner or manager of these property items then assigns rights to specific property items. Rights transactions occur when users are given specific rights to use the content. The rights management system may perform or assist in the collection of license fees or royalties. Various monitoring tools may be used to ensure authorized usage of content and to ensure revenues are collected.

Figure 9.1 shows a rights management system. This diagram shows that a rights management system oversees the identification and management of intellectual property items (content), rights assignments, rights transactions, licensing fees and usage monitoring (enforcement). This diagram shows that a rights management system oversees how content owners can provide access for content to users and how to convert and ensure the usage of content is converted into value for the content owner.

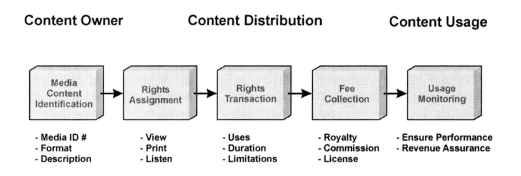

Figure 9.1, Rights Management System

The transfer of rights of intellectual property from a content owner to a content user or distributor may involve the use of a formal agreement (e.g. a publishing agreement) or it may occur through an action (e.g. a customer buying a book).

A content owner is a person or company that owns the rights to intellectual property (content). Rights users can be a person, company, or group that receives, processes or takes some form of action on services or products.

Rights may be transferred by the owner of the content or by an agent. A licensor is a company or person who authorizes specific uses or rights for the use of technology, products or services. An agent is a person or a device that performs tasks for the benefit of someone or some other device.

When rights are given for the use of content, the rights owner is called a licensee. A licensee is the holder of license that permits the user to operate a product or use a service. In the television industry, a licensee is usually the company or person who has been given permission to provide (e.g. broadcast) a particular program or service within a geographic area.

The assignment or transfer of rights may be formalized in a written rights agreement or it may occur as the result of some action such as the purchase of an item (such as the sale of a book) which transfers rights to the owner (such as the right to read, loan or destroy the book purchased). An agent may be used to assign and transfer rights. The types of rights that may be assigned include visual, audio, smell or other forms that can communicate information about intellectual property. Usage may be in the form of rendering, transferring or manipulating (changing) the intellectual property.

When rights are transferred, there is usually some form of tangible compensation defined in license terms such as licensing fees or royalties. License terms are the specific requirements and processes that must be followed as part of a licensing term agreement. Royalties are compensation for the assignment or use of intellectual property rights.

Content owners may be able to have exclusive rights to their content restricting its licensing to specific people or companies or content owners may be forced to license specific types of content to various types of users (compulsory licensing).

Compulsory licensing is the requirement imposed by a governing body that forces a holder of intellectual property (e.g. a patent) to allow others to use, make or sell a product, service or content. Compulsory licensing usually requires the user of the intellectual property (licensee) to pay the owner (licensor) a reasonable license fee along with non-discriminatory terms.

When products require the use of multiple technologies or forms of intellectual property, the owners of the intellectual property may group together to form a collective licensing system. A collective licensing system is a process that allows a collective group of technologies or intellectual property to be licensed as a complete group instead of identifying and negotiating licenses for each part separately.

The specific rights that are assigned during a rights transaction is detailed in a rights specification. Rights specification defines the ability to render (display), transport (copy and send) and derive (modify or use portions) for a specific content item.

When the transfer of rights involves the use of content, it is called a content transaction. Content transactions can range from simple one time use of content (such as viewing a movie) to the complete transfer of content rights (the sale of content rights to a publisher).

## Intellectual Property Rights

Intellectual property rights are the privileges (such as exclusive use) for the owner or the assignee of the owner of patents, trademarks, copyrights, and trade secrets.

# Copyright

A copyright is a monopoly, which may be claimed for a limited period of time by the author of an original work of literature, art, music, drama, or any other form of expression - published or unpublished. Copyrights are Intellectual Property Rights that give the owner or assignee the right to prevent others from reproducing the work or derivates including reproducing, copying, performing, or otherwise distributing the work. Copyrights in many countries can be claimed without registration. However, most countries or regions have a copyright office where copyrights can be officially registered.

Copyright protection is the legal rights and/or physical protection mechanisms (security system) that provide control of the use of an original work of literature, art, music, drama, or any other form of expression. Copyright infringement is the unauthorized use of intellectual property (copyrighted content) by a person or company.

Copyright registration is the process of recording the creation of a new work with the copyright office or intellectual property rights center in the country of origin. Registering a copyright is not required and a work may be considered copyright protected when it is initially created (such as in the United States). Copyright registration does provide additional benefits to the creator or rights holder especially when attempting to enforce copyright protection of the content. There may be a time limit between the creation of the work and registration (such as 5 years in the United States).

# Patents

A patent is a document that grants a monopoly for a limited time to an invention described and claimed in the body of the document. A patent describes the invention and defines the specific aspects (claims) of the invention that are new and unique. There are several forms of patents including mechanical patents (processes) and design patents (appearances).

To use the technologies described in patents, users must obtain patent rights. Patent rights are intellectual property rights that give the owner or assignee the right to prevent others from making, using, or selling the invention described and claimed in the patent. Patent rights must be applied for in the country or region where they are desired. Most countries have a national patent office and an increasing number of countries grant patents through a single regional patent office.

## Trademarks

A trademark is unique symbol, word, name, picture, design, or combination thereof used by firms to identify their own goods and distinguish them from the goods made or sold by others. Trademark rights are Intellectual Property Rights, which give the owner the right to prevent others from using a similar mark, which could create confusion among consumers as to the source of the goods. Trademark protection must be registered in the country or region where it is desired. In most countries and regions, patents and trademarks are administered by the same government agency.

## Trade Secrets (Confidential Information)

Trade secrets are information, data, documents, formulas, or anything which has commercial value and which is kept and maintained as confidential. An example of a trade secret is a customer list and their service or product purchase histories.

Figure 9.2 show the types of intellectual property rights that may be associated with content. This diagram shows that rights include patent, trademarks, trade secrets and copyright. Copyrights protect original works (creations) from use or performance by others. Patent rights protect the inventor (or their assignee) from the production or use of their invention by other people or companies. Trademarks protect the use of images or names that they use to promote specific products, brands or services. Trade secrets protect the exchanged of private (confidential) information from unauthorized users.

| Type of Intellectual Property | Rights Covered |
|---|---|
| Copyright | Use or performance of original works of literature, art, music, drama or any other form of expression. |
| Patent | The use, manufacture or the sale of inventions. |
| Trademark | The use of symbols, words, names, pictures, designs or combination thereof used by firms to identify particular products, brands or services. |
| Trade Secrets | The privacy of data, documents, formulas or anything that is to be maintained as confidential information. |

Figure 9.2, Types of Intellectual Property Rights

# Types of Rights

The types of rights associated with content include rendering, transporting and modifying (deriving) content.

## Rendering Rights

Rendering is process of converting media into a form that a human can view, hear or sense. An example of rendering is the conversion of a data file into an image that is displayed on a computer monitor.

Rendering rights as they apply to content include printing, displaying and playing media content. Printing is the rendering of media into a hard copy format. Displaying is the conversion of media into a display format (such as a computer or television display). Playing is the translation of media into an acoustic format.

## Transport Rights

Transport rights are the authorizations to move, copy or loan content. An example of transport rights is the transmission of television programs through a satellite system.

Transport rights as they apply to content include moving, copying and loaning content. Moving is the physical transferring of a media file from one location to another. Copying is the selection of a portion or portions of media content and replicating those selections for use or transfer to other people or companies. Loaning is the process of transferring media content to another person or company for a temporary period. Loaning content requires that the original content be removed or disabled from use.

## Derivative Rights

Derivative rights are the authorizations to extract, insert or edit content. An example of derivative rights is the use of text from a book or article in a presentation or another article.

Derivative rights as they apply to content include extracting, insertion and changing content. Extracting (using sections) is the selection and removal of content for use in another work. Inserting (embedding) is the pasting of additional content (e.g. such as a editorial comment) within an existing content item. Changing (editing) content is the modification of any content or expression of a content item.

Figure 9.3 shows basic rights management processes. This example shows that rights management can involve rendering, transferring and derivative rights. Rendering rights may include displaying, printing or listening to media. Transfer rights may include copying, moving or loaning media. Derivative rights can include extraction, inserting (embedding) or editing media.

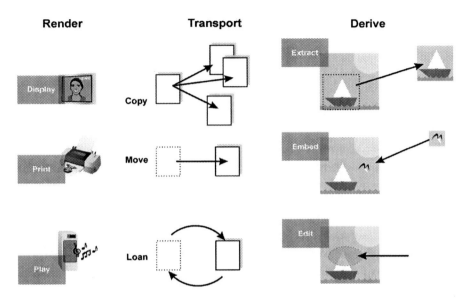

Figure 9.3, Rights Management Process

There are many other types of rights that can be associated with content. Publishing rights are the authorized uses that are granted to a publisher to produce, sell and distribute content. Publishing rights may range from global distribution in any media format to specific media formats and geographic areas. Publishing permissions are the rights granted by a publisher for the use of content that the publisher has rights to exercise.

Reprint rights are the permissions granted from an owner of content to reprint the content in specific formats (such as reprinting a magazine article).

Resale rights are the authorizations that allow a person or company to resell a product, service or content. Resale rights may specify the formats which a product may be sold (e.g. eBook) and where the product may be sold (geographic regions).

Portability rights are the permissions granted from an owner or distributor of content to transfer the content to other devices (such as from a set top box to a portable video player) and other formats (such as low bit rate versions).

## Digital Rights Management

Digital rights management is a system of access control and copy protection used to control the distribution of digital media. DRM involves the control of physical access to information, identity validation (authentication), service authorization, and media protection (encryption). DRM systems are typically incorporated or integrated with other systems such as content management system, billing systems, and royalty management. Some of the key parts of DRM systems include key management, product packaging, user rights management (URM), data encryption, product fulfillment and product monitoring.

When intellectual property is stored in a form of digital media, objects of value are called digital assets. Digital assets have rights associated with them and digital right management systems oversee how the rights are transferred between owners and users of digital assets.

Content owners who want to protect their content as it is distributed through television systems may require the use of DRM. Content owners (such as studios) may prefer to use or endorse specific types of DRM systems. A studio endorsement is the allowance of the use of a program or service as long as the use meets with specific distribution requirements. Studios may endorse (authorize) or prefer the use of specific content protection systems that can be used for the distribution of their programs and not all content protection systems meet their security requirements. Media service providers who do not use a studio endorsed content protection system may not be able to get and distribute content from some studios.

## Rights Attributes

Rights attributes are the details related with each right associated with a content item. An example of a rights attribute is the authorization to print a version of an eBook up to 3 times for a university professor. Common types of rights attributes include permissions, extents and consideration.
Permissions

Rights permission is the uses or authorizations to perform actions on or concerning the use of content. Permissions include reproduction (copying), deriving (modifying), transporting (distributing), performing or displaying (rendering) the work.

Rights Extents

Rights extents are the amount of usage of a content item that is authorized. Rights extents may be defined in units of time, the number of uses and what places (geographic regions) that the rights apply.

Rights Consideration

Rights consideration is the authorization or transfer of value in return for the assignment of rights. Rights consideration may be in the form of money or any other quantifiable items or services. Rights consideration may be a combination of royalties, commissions, grants of license fees.

## Rights Models

Rights model is a representation of how the content rights are transferred and managed between the creator or owner of content and the users or consumers of content. Examples of rights models include a radio broadcasting rights model, television-broadcasting rights model, Internet rights model and a book rights model.

## *Radio Broadcasting Rights Model*

A radio broadcasting rights model defines how the audio media content (such as music programs) is transferred to listeners (users).

The radio broadcasting rights model may start with the music content production company (e.g. recording company), which makes a production agreement with a music artist. This allows the recording company to produce and sell the music in a variety of formats in return for paying a percentage of sales (royalties) back to the music artist.

As part of the radio broadcasting rights, radio broadcasting stations sign licensing agreements with record companies or their authorized agents (associations) which allow them to broadcast the music on radio channels to listeners. The fees paid for radio broadcasting rights vary depending on the location and number of listeners the radio station has.

As part of the licensing agreement, users may listen to the music and have rights to store (record) and replay the music with certain usage restrictions. The radio broadcasting company typically receives its compensation for broadcasting from advertisers.

Figure 9.4 shows a rights model that represents music from its creation to a radio listener. This example shows how a musician signs a contract with a recording company that allows for the creation and sale of music CDs and DVDs. The recording company signs a licensing contract with a radio station that allows the music to be played on the airwaves. The listener can enjoy the music with certain usage restrictions. The radio broadcasting company pays licensing fees to the recording company, which pays a portion of this money to the musician in the form of royalties.

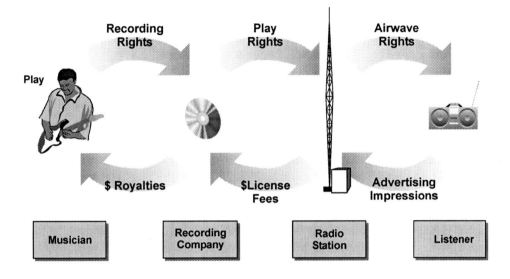

Figure 9.4, Radio Broadcast Rights Transactions

## Television Broadcasting Rights Model

A television broadcasting rights model defines how the video and audio content (such as a television program) is transferred to viewers (users).

The television broadcasting rights model may start with the program content production company (e.g. movie studio), which makes production agreements with a variety of artists (e.g. actors and directors). This allows the movie studio to produce and sell the program in a variety of formats in return for paying fees and potentially a percentage of sales (royalties) back to the artist.

As part of the television broadcasting rights, television broadcasting stations, cable television companies or other television program distributors sign licensing agreements with movie studio or their authorized agents (content aggregators) that allow them to broadcast the television programs on their distribution systems to viewers. The fees paid for television broadcasting rights vary depending on the location and number of viewers the television broadcaster has.

As part of the licensing agreement, users may view and have rights to store (record) and replay the television programs with certain usage restrictions. The television broadcasting company may receive its compensation for broadcasting from subscription (e.g. monthly subscription service fees), usage (e.g. pay per view) and advertising fees.

Figure 9.5 shows a rights model that represents television media from the creation of television shows to rights of a TV viewer. This example shows how a production studio signs contracts with actors, graphics artists and writers to product television content. Each of these content developers and providers provide broadcasting rights (and possibly other rights) to the movie studio. The production company signs a licensing agreement with a broadcast company that allows the media to be sent via television radio broadcast, cable transmission or via satellite systems. The viewer is allowed to view and store the television program for his or her own personal use. The broadcaster receives subscription and/or pay per view fees from the viewer,

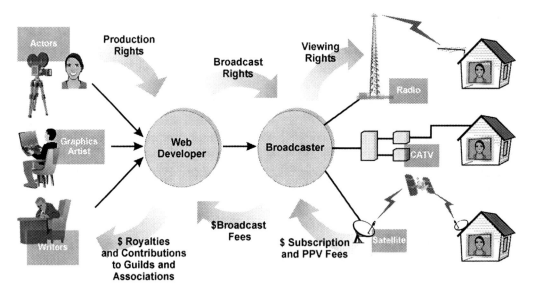

Figure 9.5, Television Rights Transactions

the production studio receives license fees from the broadcaster for distribution (possibly revenue sharing) and the contributing artists receive performance fees, royalties and potentially other contributions to artist guilds (e.g. actors guild) and associations.

## *Internet Rights Model*

An Internet rights model defines how the multimedia content (such as a web page) are transferred to Internet viewers (users).

The Internet rights model may start with a person or business that wishes to combine content into a form that can be communicated to viewers who connect to their web site through a public network (the Internet). Multiple people, companies or creators may own the content that is combined on a web site.

The person who combines the content into a format that is provided by the web site is called a webmaster. The webmaster is usually responsible for ensuring that only authorized information is stored and provided on the web site.

As part of the licensing agreement, users may view and have rights to store (save) and transfer the media captured from the web site. The company that manages the content on the web site may make money through the sale of products, reduced cost of supporting product sales (e.g. product help screens) or through the insertion and viewing of advertising messages (e.g. banner ads).

Figure 9.6 shows a rights model that represents how content may be created and hosted on the Internet and access by a web user. This example shows how a web developer obtains content or developer agreements with videographers, photographers and recording artist to provide media components. Each of these content providers and developers provide web casting rights (and possibly all other rights) to the web site developer. The web development signs an agreement with a web hosting company which requires the web developer to have all the rights necessary to transfer the content to users. Web users obtain user rights when they connect to the web host and transfer media to their computer. This example shows that these rights may include viewing, printing, forwarding, and storing the media on their personal computer. The web host receives compensation from the web user in the form of page visits and/or link clicks (which may be paid for by another company). The web developer then receives development fees from the web host and the content providers receive licensing fees and/or royalties from the web developer.

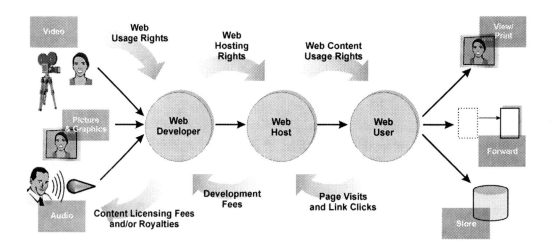

Figure 9.6, Internet Rights Transactions

## Digital Assets

A digital asset is a digital file or data that represents a valuable form of media. Digital assets may be in the form of media files, software (e.g. applications) or information content (e.g. media programs).

Digital content is raw media (essence) that is combined with descriptive elements (metadata). Metadata is information (data) that describes the attributes of other data. Metadata or Meta tags are commonly used to provide technical and descriptive information about other data. (Metadata is data that describes other data). A television program may contain technical metadata that describes the media encoding formats and duration of the media along with descriptive metadata that provides information about the title and actors.

In addition to digital assets that may be acquired or originally recorded (such as video from a sports game), digital assets may be combined to create new digital assets through the production and editing process. This process produces secondary content. An example of secondary content is the creation of "sports highlights" from a series of games that were previously broadcasted.

Digital assets are typically stored in a high-resolution version so they that can be converted (transcoded) into lower resolution formats as required. High-resolution digital assets require an extensive amount of storage area so older or less used digital assets are commonly moved to offline repositories. The original digital assets (master files) are commonly stored in a secure area (called a "vault") which usually has restrictions on use or have usage costs associated with the content. This vault may be an electronic storage facility (e.g. videodisks) that is protected by a password system and data network firewalls.

Some of the key types of digital assets include digital audio, images, video, animation and application programs. There can be many other types of digital assets including web pages, scripted programs (e.g. Javascript files), fonts and other types of media that can be represented in digital form.

## Digital Audio (Voice and Music)

Digital audio is the representation of audio information in digital (discrete level) formats. The use of digital audio allows for more simple storage, processing, and transmission of audio signals.

The creation of digital audio typically involves the conversion of analog signals into digital formats (it is possible to create synthetic audio without analog signals). The conversion process determines the fidelity of the digital audio signal. Audio fidelity is the degree to which a system or a portion of a system accurately reproduces at its output the essential characteristics of the signal impressed upon its input.

The typical data transmission rate for uncompressed high quality audio is 1.5 Mbps (for 2 channels of audio). This requires a data storage area of approximately 11 MBytes for each minute of storage. Digital audio is commonly compressed (coded) to reduce the amount of storage that is required for the audio. There key ways that digital audio may be coded including waveform coding, media or perceptual coding. Broadcast audio is commonly compressed and transmitted in an MPEG format (e.g. MP3) at 128 kbps.

Waveform coding consists of an analog to digital converter and data compression circuit that converts analog waveform signal into digital signals that represent the waveform shapes. Waveform coders are capable of compressing and decompressing voice audio and other complex signals.

Media coding is the process of converting an analog waveform into components that represent its underlying media format. For example, speech coding analyzes the analog signal into codes that represent human sounds. Media coders typically provide a substantial amount of compression (8:1 ore more). However, media coders are only capable of transforming signals of the media types they are designed for. This is why music does not typically sound good through mobile telephones as mobile telephones compress human audio and they are not designed to compress the sounds of musical instruments.

Perceptual coding is the process of converting information into a format that matches the human senses ability to perceive or capture the information. Perceptual coding can take advantage of the inability of human senses to capture specific types of information. For example, the human ear cannot simultaneously hear loud sounds at one tone (frequency) and soft sounds at another tone (different frequency). Using perceptual coding, it is not necessary to send signals that cannot be heard even if the original signal contained multiple signals. MP3 coding format uses perceptual coding.

## Images

Images are data files that organize their digital information in a format that can be used to recreate a graphic image when the image file is received and decoded by an appropriate graphics application. Images may be used within or added to (e.g. logos) program materials.

The creation of digital images may involve the conversion of pictures or images into digital format (e.g. scanning or digital photographing) or it may be through the creation of digital components (image editing).

The number of pixels per unit of area determines the resolution of the image. A display with a finer grid contains more pixels, therefore it has a higher resolution capable of reproducing more detail in an image. The raw file size is determined by how many bits represent each pixel of information. A single image that has high resolution can require over 200 MBytes per image (600 dpi x 11" horizontal x 600 dpi x 17" vertical x 24 bits (3 Bytes) per pixel = 200+ MBytes).

## Animation

Animation is a process that changes parameters or features of an image or object over time. Animation can be a change in position of an image within a video frame to synthetically created images that change as a result of programming commands.

Animation media is the set of data and images along with their controlling software commands that are used to produce animated displays. The production of animated images may be performed at the broadcaster's facility or they may be performed at the viewer's location (e.g. gaming applications).

## Digital Video

Digital video is a sequence of picture signals (frames) that are represented by binary data (bits) that describe a finite set of color and luminance levels. Sending a digital video picture involves the conversion of a scanned image to digital information that is transferred to a digital video receiver. The digital information contains characteristics of the video signal and the position (or relative position) of the image (bit location) that will be displayed.

The typical data transmission rate for uncompressed standard definition (SD) television is 270 Mbps and uncompressed high definition (HD) digital video is approximately 1.5 Gbps. This requires a data storage area of approximately 200 MBytes for each minute of standard definition storage and more than 11 GBytes for 1 minute of high definition television storage.

Digital video is commonly compressed (coded) to reduce the amount of storage that is required for the video. The ways that digital video may be coded (compressed) including spatial compression and temporal compression. Spatial compression (within a frame) is the analysis and compression of information or data within a single frame, image or section of information. Temporal compression (between frames) is the analysis and compression of information or data over a sequence of frames, images or sections of information.

Combining spatial compression and temporal compress, the data transmission rate for digital video can be reduced by 200:1 or more. Broadcast video is commonly compressed and transmitted in an MPEG format at 2 to 4 Mbps (2 hours of video is approximately 4 GBytes = 1 DVD). During the editing and production process, the use of compression is minimized to ensure the quality of the original materials.

## Data Files (Books and Databases)

A data file is a group of information elements (such as digital bits) that are formatted (structured) in such as way to represent information. While the term data file is commonly associated with databases (tables of data), a data file can contain any type of information such as text files or electronic books.

## Application Program Files (Software Applications)

An application program is a data file or group of data files that are designed to perform operations using commands or information from other sources (such as a user at a keyboard). Popular applications that involve human interface include electronic mail programs, word processing programs and spreadsheets. Some applications (such as embedded program applications) do not involve regular human interaction such as automotive ignition control systems.

## Container Files

A container is a structured format of a media file that can hold a group of items. A container may hold descriptors (metadata), indexes and/or raw media objects (e.g. digital video). Contain files hold items that contain raw data (essence) and descriptions (metadata). Items within a container file may contain other items.

Figure 9.7 shows how a file container can hold multiple media items (components) and how each item has a description. This diagram shows a container that is composed of 3 main items and one of the items contains an item within itself. This example shows that each item has a descriptor that provides details about its components. Each component also has a descriptor that provides specific information about its resources (e.g. raw media).

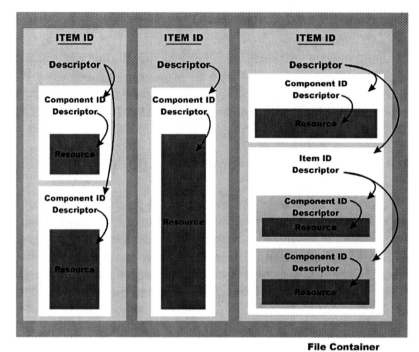

Figure 9.7, Media file Container

There are various container file formats that are used to hold content for the television industry. Editing and production may use the advanced authoring format (AAF) to hold digital media (essence and metadata). In addition to holding the raw media in a format that can be used by production people, the AAF structure includes additional information that can assist in the production (authoring) process.

When creating content segments for broadcast programs, the material exchange format (MXF) format may be used. MXF is a limited version (subset) of the AAF structure. The metadata within the MXF media format describes how the media elements (e.g. video and audio) are re-assembled into their original media format.

# Digital Asset Management (DAM)

Digital asset management is the process of acquiring (ingesting), maintaining (managing assets), and distributing (broadcasting and transferring) digital assets.

## *Ingestion*

Ingesting content is a process for which content is acquired and loaded onto initial video servers (ingest servers). Content ingestion may come from several different sources including satellite downlinks, stored media (DVDs, tapes) or data connections.

While ingestion may be directly controlled from media players (such as DVD players), some content is transferred via data networks and there is a cost associated with the transfer of information through these networks. In general, as the data connection speed and distance increases, the cost of the data connection dramatically increases. As a result, some media may be ingested slowly to allow for the user of slower, low cost data connections.

## *Cataloging*

Cataloging is the process of identifying media and selecting groups of items to form a catalog or index of the stored media. As media is ingested into a DAM system, it is identified (labeled) along with key characteristics and added to the catalog of materials. Cataloging is a key part of the asset manager.

The first step in cataloging is labeling (logging) the digital asset when it is acquired or created. After the digital asset is identified, additional information is associated with the digital asset. This information may be technical or descriptive data. Much of this information may already be available within the metadata portion of the asset that is ingested.

The cataloging process is becoming more automated and much of the metadata that is already provided with the media may be automatically imported, categorized and stored with the digital asset. To ensure that descriptions are clear and compatible between networks (e.g. not to confuse the category of thriller with pornography), metadata industry standards such as MPEG-7 are being developed.

## Digital Asset Storage

Digital asset storage is a process of transferring data into or onto a storage medium (a repository) in which information can be retained. Some of the common types of digital storage mediums include electronic (e.g. RAM), magnetic (e.g. tape or hard disk) and optical (e.g. CDROM or DVD).

Repositories may be immediately accessible (such as video servers) or the storage media may need to be selected and loaded (such as videodisks). Assets that are directly accessible to users and devices connected to the media system are called online assets.

Because direct and fast online storage is relatively expensive, digital assets may be stored in remote nearline or offline storage systems that require some time to find, select and prepare access to the media. Nearline assets are files or data that represents a valuable form of media that are not immediately available for use but can be made available for use in a short period of time. Offline assets are files or data that represents a valuable form of media that is not directly or immediately available for use.

## Editing and Production

Editing and production is the searching, selection, connection, and combining of media materials to produce programs or contents segments. To create a program, program editors identify content segments to use and the starting and ending points in these segments by developing and edit decision list (EDL).

An EDL is a set of records that contain the names of media files and time position locations within these files (edit in-points and out-points) that are used to create a video program. The media that is referenced in an EDL list does not have to be available during its creation. The EDL may be created using an off-line editing system. The media only needs to be accessible when the EDL is processed to create the final master video program.

## *Playout and Distribution*

Once content is ingested it can be edited to add commercials, migrated to a playout server or played directly into the transmission chain. Playout is the process of streaming or transferring media to a user or distributor of the media. Playout may be in the form of real time broadcasting, through scheduled or on demand delivery.

The distribution of content may be in real time, scheduled or on demand delivery. Real time broadcasting involves directly converting (transcoding) media into a format that can be transferred through a distribution network such as a direct to home (DTH) satellite or cable television system. Prerecorded media may be scheduled for delivery to networks or other distribution systems. This delivery schedule may occur several hours before the program is scheduled to be broadcasted to customers. Scheduled programs may be transmitted at a slower transmission rate to reduce the costs of data transmission. Some asset management systems allow broadcasters to connect to the asset management system and download the materials when they require them (or when they can cost effectively transfer them).

Figure 9.8 shows a digital asset management system. This diagrams shows that the basic functions for a digital asset management system include ingestion, cataloging and distribution. This example shows that content may be ingested from a variety of sources including live feeds or analog and digital stored media formats. New content may also be created through the editing and production of existing assets. The asset manager coordinates the storage and retrieval of media into online (cache), nearline and offline repositories and coordinating the conversion (transcoding) of media into other formats (such as MPEG). This example shows that the asset manager creates and uses catalog information to allow the system and users to find, organize and obtain media. The distribution portion is responsible for transferring assets to broadcast (real time), other storage systems (time delayed) and/or to Internet streams

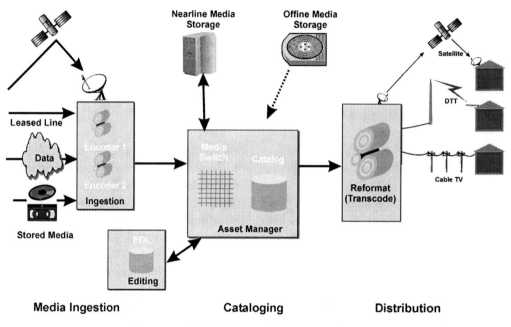

Figure 9.8, Digital Asset Management System

# Media Identification

Digital media content must be uniquely identified. Some of the common identification codes that may be assigned to digital media include UMID, ISBN (books), ISSN (magazines), DOI (multimedia) and watermarks.

### Unique Material Identifier (UMID)

A unique material identifier is a unique code that is used to identify audio-visual (AV) materials. The UMID standard was adopted as SMPTE standard 330M. The basic UMID contains 32 bytes of information and the first 12 bytes are used as a label to indicate that the media has a UMID code. Bytes 13 through 15 define the length of the material and the last 16 bytes of the UMID are a unique number that identifies the specific content item.

A UMID can be extended with another 32 bytes of content descriptive information to create an extended UMID (EUMID). The EUMID contains the date, time and location where the content originated.

Figure 9.9 shows the structure of an EUMID identifier. This diagram shows that an EUMID is composed of the basic 32 byte UMID that contains the media identification information plus additional 32 bytes of descriptive information. This additional descriptive information includes the data and time the media was created, the place (location) that the media was created, along with some additional user information.

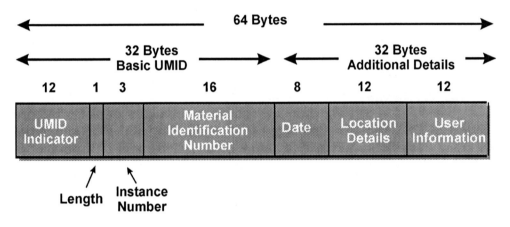

Figure 9.9, EUMID Identifier

## International Standard Book Number (ISBN)

An international standard book number is a unique number assigned to a specific book. The ISBN identifies the issuing group, language or geographic area, publisher, title identifier, and a calculated check digit to ensure an ISBN has been entered correctly into a computer or system.

## International Standard Serial Number (ISSN)

An international standard serial number is a unique number that identifies the particular issue of a series of magazines or periodical publications. The ISSN number contains a unique portion that identifies the particular series of publication and a check digit to ensure an ISBN has been entered correctly into a computer or system. While the ISSN identifies the series publication, the particular issue (e.g. volume or issue) has the series idefined by the publisher and it is not part of the ISSN.

## Digital Object Identifier (DOI)

A digital object identifier is a unique number that can be used to identify any type or portion of content. DOI numbers perform for long term (persistent) and locatable (actionable) identification information for specific content or elements of content. This content can be in the form of bar codes (price codes), book or magazine identification numbers or software programs. The DOI system is managed by the International DOI foundation (IDF) that was established in 1998. More information about DOI numbering can be found at www.DOI.org.

DOI numbers point to a DOI directory that is linked to specific information about a particular object or its information elements. The use of a DOI directory as a locating mechanism allows for the redirecting of information about identification information as changes occur in its identifying characteristics. For example, a book identification number may belong to the original publisher until the copyright of the work is sold to another publisher. At this time, the owners of the item content changes. The item number on the book can remain the same while the publisher information can change. Once a DOI is created, it is considered permanent and is not deleted.

The numbering structure (syntax) for a DOI includes a directory and owner part (prefix) and unique item identification number (suffix). The prefix allows the DOI directory to redirect (DOI resolution) identification information to the owner or controller of the identifiable item. This redirection may be to a URL web link.

While it is important that there be only one DOI assigned per content object, a single object may have different types of identification numbers assigned to it. For example, a book can have a DOI, international standard book number (ISBN) and a manufacturers product number assigned to it.

Figure 9.10 shows digital object identifier (DOI) structure. This example shows that a DOI number is composed of a prefix that is assigned by the registration agency (RA) of the international DOI foundation and suffix that is assigned by the publisher of the content. This example shows that the first part of the prefix identifies the DOI directory that will be used and the second part identifies the publisher of the content.

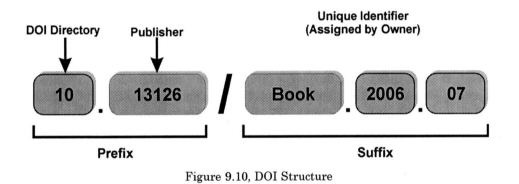

Figure 9.10, DOI Structure

## International Standard Recording Code (ISRC)

The international standard recording code is a unique code that is approved by the International Standard Organization (ISO) to identify audio-visual (AV) recorded materials. Each ISRC code is unique and permanently stored within the recorded media program.

## International Standard Audiovisual Number (ISAN)

International standard audiovisual number is a unique code that is approved by the International Standard Organization (ISO) in standard

15706 to identify AV materials or portions of AV materials. The International ISAN Agency maintains and coordinates a centralized database of all ISAN records.

### International Standard Work Code (ISWC)

International standard work code is a unique code that is approved by the International Standard Organization (ISO) in standard 15707 to identify materials or portions of materials for the music industry. The International ISWC Agency maintains and coordinates a centralized database of all ISWC records.

## DRM Security Processes

Security is the ability of a person, system or service to maintain its desired well being or operation without damage, theft or compromise of its resources from unwanted people or events. DRM systems are designed to protect the rights of the media owners while enabling access and distribution of the media programs to users.

DRM systems are designed to establish and maintain security associations between two network elements that ensures that traffic passing through the interface is cryptographically secure (typically, through the use of encryption). DRM systems can use a combination of digital watermarks, digital fingerprints, digital certificates, digital signatures, conditional access systems, product activation codes, authentication and encryption to provide security assurances to media content and their delivery systems.

### Authentication

Authentication is a process of exchanging information between a communications device (typically a user device such as a mobile phone or computing device) and a communications network that allows the carrier or network operator to confirm the true identity of the user (or device). This validation of the authenticity of the user or device allows a service provider to deny service to users that cannot be identified. Thus, authentication inhibits fraudulent use of a communication device that does not contain the proper identification information.

Authentication credentials are the information elements that are used to identify and validate the identity of a person, company or device. Authentication credentials may include identification codes, service access codes and secret keys.

A common way to create identification codes that cannot be decoded is through the use of password hashing. Password hashing is a computational processes that converting a password or information element into a fixed length code. Password hashing is a one way encryption process as it is not possible to derive the original password from the hashed code.

Figure 9.11 shows how hashing security can be used to create identification codes that cannot be directly converted back into the original form. In this example, a hash code is created from the password 542678. The hashing

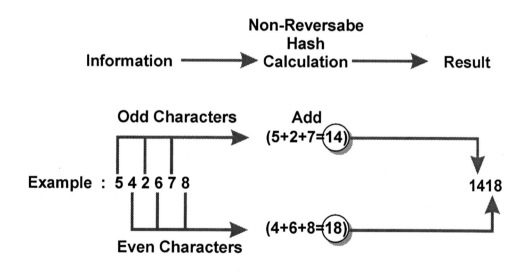

Figure 9.11, Password Hashing

process involves adding the odd digits to calculate the result (14), adding the even digits to calculate a result (18) and then storing the results of the password (1418). By using this hashing process, the original password cannot be recreated from the hashing results, even if you know the hashing process.

Figure 9.12 shows a typical authentication process used in a DRM system. In this diagram, a DRM server wants to validate the identity of a user. The DRM system has previously sent a secret key to the user. The authentication process begins with the DRM server sending an authentication request and a random number. This random number is used by the receiving device and is processed with the secret key with an authentication (data processing) algorithm to produce a calculated result. This result is sent to the originator (authenticator). The originator uses the random number it sent along with its secret key to calculate a result. If the result received from the remote device with its own result matches, the authentication passes. Note that the secret key is not sent through the communication network and that the result will change each time the random number changes.

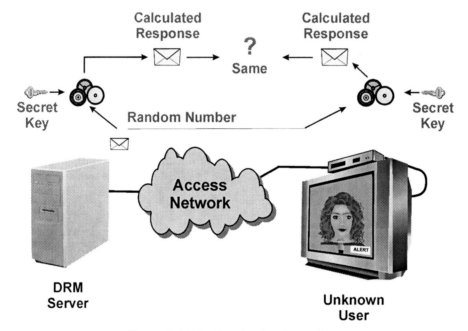

Figure 9.12, Authentication Operation

## Encryption

Encryption is a process of a protecting voice or data information from being used or interpreted by unauthorized recipients. Encryption involves the use of a data processing algorithm (formula program) that uses one or more secret keys that both the sender and receiver of the information use to encrypt and decrypt the information. Without the encryption algorithm and key(s), unauthorized listeners cannot decode the message.

Figure 9.13 shows the basic process used by encryption to modify data to an unrecognizable form. In this example, the letters in the original information are shifted up by 1 letter (example - the letter I becomes the letter J). With this simple encryption, this example shows that the original information becomes unrecognizable by the typical viewer.

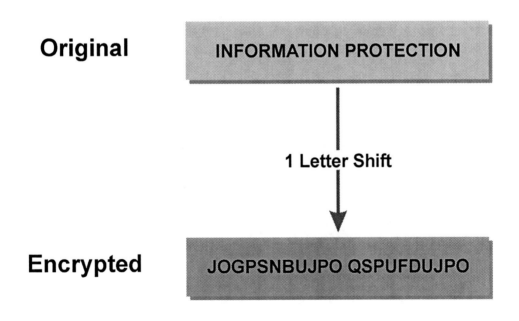

Figure 9.13, Basic Encryption Process

## Encryption Keys

An encryption system typically uses a combination of a key (or keys) and encryption process (algorithm) to modify the data. An encryption key is a unique code that is used to modify (encrypt) data to protect it from unauthorized access. An encryption key is generally kept private or secret from other users. Encryption systems may use the same encryption key to encrypt and decrypt information (symmetrical encryption) or the system may use different keys to encrypt and decrypt (asymmetrical encryption). Information that has not been encrypted is called cleartext and information that has been encrypted is called ciphertext.

The encryption key length is the number of digits or information elements (such as digital bits) that are used in an encryption (data privacy protection) process. Generally, the longer the key length, the stronger the encryption protection.

Figure 9.14 shows how encryption can convert non-secure information (cleartext) into a format (cyphertext) that is difficult or impossible for a

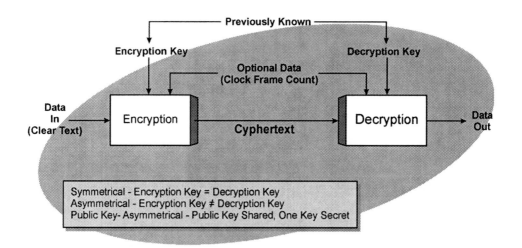

Figure 9.14, Encryption Operation

recipient to understand without the proper decoding keys. In this example, data is provided to an encryption processing assembly that modifies the data signal using an encryption key. This diagram also shows that additional (optional) information such as a frame count or random number may be used along with the encryption key to provide better information encryption protection.

## Symmetric and Asymmetric Encryption

Encryption systems may use the same key for encryption and decryption (symmetric encryption) or different keys (asymmetric encryption). Generally, asymmetric encryption requires more data processing than symmetric encryption.

Figure 9.15 shows the differences between symmetric and asymmetric encryption. This diagram shows that symmetrical encryption uses the same keys to encrypt and decrypt data and that asymmetric encryption uses different keys and processes to encrypt and decrypt the data.

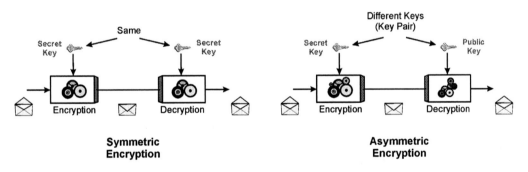

Figure 9.15, Symmetric and Asymmetric Encryption Processes

## Public Key Encryption

The encryption process may be private or public. Public key encryption is an asymmetric authentication and encryption process that uses two keys, a public key and a private key, to setup and perform encryption between communication devices. The public key and private keys can be combined to increase the key length provider and more secure encryption system. The public key is a cryptographic key that is used to decode information that was encrypted with its associated private key. The public key can be made available to other people and the owner of the key pair only uses the private key.

The encryption process may be continuous or it may be based on specific sections or blocks of data. A media block is a portion of a media file or stream that has specific rules or processes (e.g. encryption) applied to it. The use of blocked encryption may make re-synchronization easier which may occur if there are transmission errors.

There are several types of encryption processes used in DRM systems including PGP, RSA, DES, AES, RC5 and ISMACrypt.

## Pretty Good Privacy (PGP)

Pretty good privacy is an open source public-key encryption and certificate program that is used to provide enhanced security for data communication. It was originally written by Phil Zimmermann and it uses Diffie-Hellman public-key algorithms.

## Rivest, Shamir and Aldeman (RSA)

Rivest, Shamir and Adleman is a public key encryption algorithm named after its three inventors. The RSA algorithm is an encryption process that is owned by RSA Security, Inc. The RSA algorithm was patented in 1983 (patent number 4,405,829).

The RSA encryption process encodes the data by raising the number to a predetermined power that is associated with the receiver, divided the number by predetermined prime numbers that are also associated with the receiver and transferring the remainder (residue) as the ciphertext.

## Data Encryption Standard (DES)

The data encryption standard is an encryption algorithm that is available in the public domain and was accepted as a federal standard in 1976. It encrypts information in 16 stages of substitutions, transpositions and non-linear mathematical operations.

Triple data encryption standard (3DES) is a variation of the data encryption standard that adds complexity to the encryption process by increasing the difficulty to break the encryption process.

## Advanced Encryption Standard (AES)

Advanced encryption standard (AES) is a block data encryption standard promoted by the United States government and based on the Rijndael encryption algorithm. The AES system uses a fixed block size of 128 bits and can have key sizes of; 128, 192 or 256 bits. The AES standard is supposed to replace the Data Encryption Standard (DES).

## Rivest Cipher (RC5)

RC5 is a symmetric block encryption algorithm that was developed by the expert cryptographer Ronald L. Rivast. RC5 is a relatively simple, efficient encryption process that can use variable block sizes and key lengths. The RC5 encryption process has block sizes of 32, 64 or 128 bits and can use a key size that can range from 0 to 2040 bits.

## International Streaming Media Association (ISMACrypt)

International streaming media association encryption is a privacy coding process that was developed by the ISMA for streaming applications.

## Digital Watermarks

A digital watermark is a signal or code that is hidden (typically is imperceptible to the user) in a digital signal (such as in the digital audio or a digital image portion) that contains identifying information. Ideally a digital watermark would not be destroyed (that is, the signal altered so that the hidden information could no longer be determined) by any imperceptible processing of the overall signal. For example, a digital watermark should not be distorted or lost when the signal is passed through a conversion or compression process.

A software program or assembly that can separate the watermark from a media file extracts a digital watermark. This watermark may be used to provide the key that is able to decode and play the media file. The process of watermarking is called stenography.

Watermarks can be encrypted to increase the resistance of the DRM system to hackers. Encrypted watermarks are tamper resistant information that is added (data embedding) or changed information in a file or other form of media that can be used to identify that the media is authentic or to provide other information about the media such as its creator or authorized usage. While it may be possible to identify the watermark in the media file, a decryption code is needed to decipher the contents of the watermark message.

Digital watermarks can be added to any type of media files such as digital video and audio. Adding or slightly modifying the colors and/or light intensities in the video in such a way that the viewer does not notice the watermarking information may perform video watermarking. Audio watermarking may be performed by adding audio tones above the normal frequency or by modifying the frequencies and volume level of the audio in such a way that the listener does not notice the watermarking information.

Figure 9.16 shows how watermarks can be added to a variety of media types to provide identification information. This example shows that digital watermarks can be added to digital audio or video media by making minor

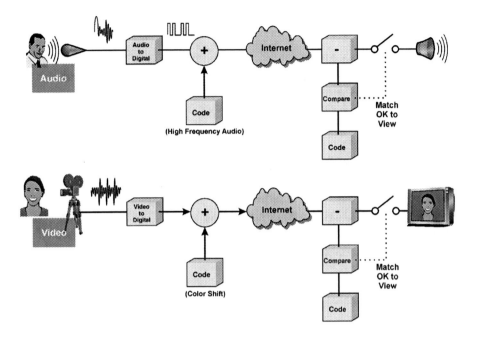

Figure 9.16, Digital Watermarking Operation

changes to the media content. The digital watermark is added as a code that is typically not perceivable to the listener or viewer of the media. This example shows that digital watermarks can be added to audio signals in the form of audio components (e.g. high frequency sound) or video components (color shift) that cannot be perceived by the listener or viewer.

## Digital Fingerprint

A digital fingerprint is a unique set of characteristics and data that is associated with a particular data file, transmission system or storage medium. Digital fingerprints may be codes that are uniquely embedded in a media file or they may be unique characteristics that can be identified in the storage or transmission medium such as the particular variance of digital bits that are stored on a DVD.

# Digital Certificate

A digital certificate is information that is encapsulated in a file or media stream that identifies that a specific person or device has originated the media. Certificates are usually created or validated by a trusted third party that guarantees or assures that the information contained within the certificate is valid.

A trusted third party is a person or company that is recognized by two (or more) parties to a transaction (such as an online) as a credible or reliable entity who will ensure a transaction or process is performed as both parties have agreed. Trusted third parties that issue digital certificates are called a certificate authority (CA). The CA typically requires specific types of information to be exchanged with each party to validate their identity before issuing a certificate.

The CA maintains records of the certificates that it has issued in repositories and these records allow the real time validation of certificates. If the certificate information is compromised, the certificate can be revoked.

Figure 9.17 shows how digital certificates can be used to validate the identity of a provider of content. This diagram shows that users of digital certificates have a common trusted bond with a certificate authority (CA). This diagram shows that because the content owner and content user both exchange identification information with the CA, they have an implied trusted relationship with each other. The content user registers with the CA and receives a certificate from the CA. The content owner registers with the CA and receives a key pair and a certificate signed by the CA. When the user requests information from a content owner, the content owner sends their public key that is in the signed certificate. Because the user can validate the signature on the certificate using the CA's public key, the user can trust the certificate and use the public key provided by the content owner (such as an online store).

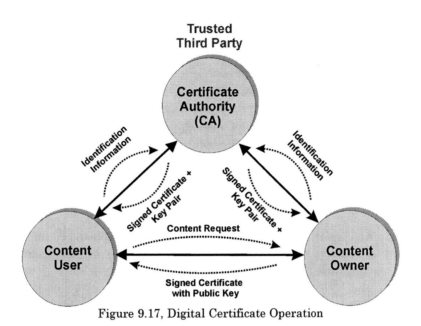

Figure 9.17, Digital Certificate Operation

## Digital Signature

A digital signature is a number that is calculated from the contents of a file or message using a private key and appended or embedded within the file or message. The inclusion of a digital signature allows a recipient to check the validity of file or data by decoding the signature to verify the identity of the sender.

To create a digital signature, the media file is processed using a certificate or validated identifying information using a known encoding (encryption) process. This produces a unique key that could only have been created using the original media file and identifying certificate. The media file and the signature are sent to the recipient who separates the signature from the media file and decodes the key using the known decoding (decryption) process.

## Secure Hypertext Transfer Protocol (S-HTTP)

Secure hypertext transfer protocol is a secure version of HTTP protocol that is used to transmit hypertext documents through the Internet. It controls and manages communications between a Web browser and a Web server. S-HTTP is designed to privately send and receive messages without the need to setup and maintain a security session.

## Machine Binding

Machine binding is the process of linking media or programs to unique information that is located within a computer or machine so the media or programs can only be used by that machine.

Some of these characteristics may include the combination of a processor type, the date the operating system was first installed and the memory storage capacity of other key characteristics that are unlikely to change over time.

## Conditional Access (CA)

Conditional access is a control process that is used in a communication system (such as a broadcast television system) that limits the access of media or services to authorized users. Conditional access systems can use uniquely identifiable devices (sealed with serial numbers) and may use smart cards to store and access secret codes. CA systems use a subscriber management system coordinates the additions, changes, and terminations of subscribers of a service.

## Product Activation

Product activation is the process of enabling a product to begin operation by entering information into the product through the use of either local operations (e.g. user keypad) or via an external connection (downloading the information into the product). Product activation usually requires that certain customer financial criteria must also be met before the product activation is performed or before the necessary product entry information is provided to the customer.

## Secure Socket Layer (SSL)

A secure socket layer (SSL) is a security protocol that is used to protect/encrypt information that is sent between end user (client) and a server so that eavesdroppers (such as sniffers on a router) cannot understand (cannot decode) the data. SSL version 2 provides security by allowing applications to encrypt data that goes from a client, such as a Web browser, to a matching server (encrypting your data means converting it to a secret code). SSL version 3 allows the server to authenticate (validate the authenticity) the client.

SSL uses a public asymmetric encryption process that uses a key pair. One key is private and one key is public. The owner of the key pair only uses the private key and the public key can be shared with others. The SSL process uses certificates from trusted third parties (certificate authorities) that validate the authenticity of messages and users.

Because the asymmetric encryption process used by the SSL system is relatively complex and time consuming to process, the SSL system may change its encryption process to use symmetric encryption after the initial asymmetric public key secure link has been setup.

Figure 9.18 shows how secure socket layer can be used to protect the transfer of digital information between the user (client) and the provider (server) of the information. This diagram shows that SSL operation uses asymmetric public key encryption to allow the sharing of public keys and that a certificate authority is used as a trusted third party to provide public keys that are accessible and verifiable. In this example, the CA has provides a key pair (public and private key) to a vendor. The server (vendor) provides their public key in their certificate, which allows the user to decode encrypted messages that are provided by the vendor. The vendor certificate is signed

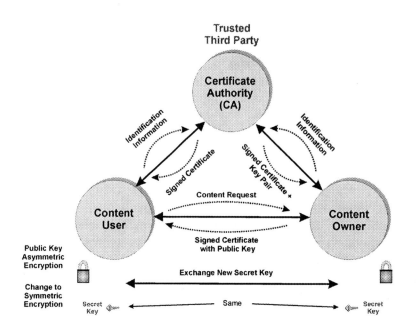

Figure 9.18, Secure Socket Layer (SSL) Operation

by the CA, which can be verified by the user. This example shows that after SSL public key encryption link is established, the SSL system can exchange keys that can be used for a less complex symmetrical encryption process.

## Transport Layer Security (TLS)

Transport layer security is a set of commands and processes that are used to ensure data transferred across a connection is private. TLS is composed of two protocol layers; TLS record protocol and TLS handshake protocol. TLS is the evolution of secure socket layer (SSL) protocol that was developed by Netscape. TLS is defined in RFC 4346.

# DRM System

DRM systems are a combination of processes, programs and hardware that facilitate the assignment, management and enforcement of the rights associated with digital media. DRM systems must identify and describe media, assign and find rights associated with that media and ensure requests for media come from authorized users,

Media communication systems are usually composed of unknown (untrusted) and known (trusted) devices. Untrusted devices are hardware components or software applications that are unknown or not validated with a provider of data or information. An untrusted device may require authentication based on some type of user interaction before access is granted. A trusted device is a product or service that is previously known or suspected to only communicate information that will not alter or damage equipment of stored data. Trusted devices are usually allowed privilege levels that could allow data manipulation and or deletion.

DRM systems are typically setup as client server systems where the system receives requests and provides services to clients. A DRM system server may include the content server (the content source), content descriptions (metadata), DRM packager (media formatter), license server (rights management), and a DRM controller (DRM message coordinator). A DRM client typically includes a DRM controller, security interface (key manager) and a media decoder.

Figure 9.19 shows a sample network architecture that could be used for a DRM system. This diagram shows a client that is requesting content and that a server processes these requests. When the user requests access to content, the DRM server first validates the identity (authenticates) of the user by using shared secret information that the client should possess. The DRM system then reviews the authorization of the user for this content and assigns rights for this use of content. A key server assigns and transfers keys that are used to encode and decode media and control messages. The license server may initiate an encryption code that is used to process the media. The DRM packager uses the encryption code to process the content and its associated descriptive (metadata) content. When the encrypted file is received in the client, the security interface may be used to gather the necessary keys and parameters necessary to decode the encrypted information.

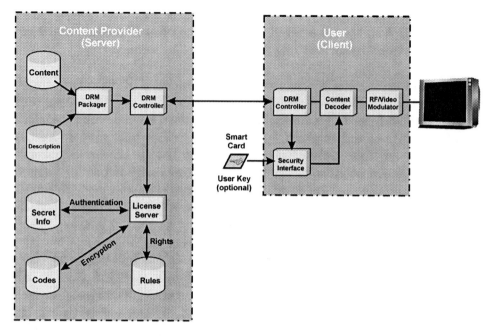

Figure 9.19, DRM Architecture

## Content Server

A content server is a computer system that provides content or media to devices that are connected to a communication system (such as through a television system). The content servers' many functions are to process requests for media content setup a connection to the requesting device and to manage media transfer during the communication session. Content servers are typically protected from direct connection to access devices by firewalls or media packing devices.

## Metadata

Metadata is information (data) that describes the attributes of other data. Metadata or meta-tags are commonly associated in media files or programs to describe the attributes of the media content. These attributes typically include the title of the media, media format details (media length and encoding formats) and may include additional descriptive information such as media category, actors or related programs.

## DRM Packager

A DRM packager is a program or system that is used to combine content (digital audio and/or video), product information (e.g. Metadata) and security codes to a media format or file that is sent from a content provider to a user or viewer of the content.

## License Server

A license server is a computer system that maintains a list of license holders and their associated permissions to access licensed content. The main function of a license server is to confirm or provide the necessary codes or information elements to users or systems with the ability to provide access to licensed content. The license server may download a key or other information to client devices that enables a license holder to access the information they have requested.

License servers use licensing rules to determine the users or devices that have authorization to access data or media. Licensing rules are the processes and/or restrictions that are to be followed as part of a licensing agreement. Licensing rules may be entered into a digital rights management (DRM) system to allow for the automatic provisioning (enabling) of services and transfers of content.

## Key Server

A key server is a computer that can create, manage, and assign key values for an encryption system. A key is a word, algorithm, or program that used to encrypt and decrypt a message that is created in a way that does not allow a person or system to discover the process used to create the keys.

DRM systems may have the capability to transfer and update keys (key renewability). Key renewability is the ability of an encryption system to issue new keys that can be used in the encoding or decoding of information.

## DRM Controller

A DRM controller is the coordinator of software and/or hardware that allows users to access content through a digital rights management system. DRM controllers receive requests to access digital content, obtain the necessary information elements (e.g. user ID and key codes), performs authentication (if requested) and retrieves the necessary encryption keys that allows for the decoding of digital media (if the media is encoded).

## DRM Client

A digital rights management client is an assembly, hardware device or software program that is configured to request DRM services from a network. An example of a DRM client is a software program (module) that is installed (loaded) into a converter box (e.g. set top box) that can request and validate information between the system and the device in which the software is installed.

DRM clients may communicate with a module (such as a smart card) or external device to manage keys and decryption programs.

# Media Transfer

Media transfer is the process of moving media from the content owner or content distributor to the end user or through other distribution systems. Media transfer may be in the form of file downloading, media streaming or through the use of stored media.

## File Downloading

File downloading is the transfer of a program or of data from a computer storage device (such as a web Internet server) server to another data storage or processing device (such as a personal computer). File download commonly refers to retrieving files from a remote server to another device or computer. File downloading is primarily used for non-real time applications such as when a program or application is completely downloaded before it is played or used.

Figure 9.20 shows how file downloading can be used to download movies through the Internet before they are viewed. This diagram shows how the media server transfers the entire media file to the media player before viewing begins. Because the media file has been completely downloaded before playing, this allows uninterrupted viewing and navigation (e.g. play, stop, rewind or fast forward).

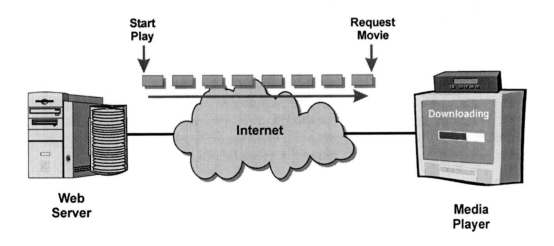

Figure 9.20, Movie Downloading Operation

## Media Streaming

Media streaming is a real-time system for delivering digital media through a data network (such as an IP data network). Upon requesting a program, a server system will deliver a stream of media (such as audio and or video) to the requesting client device (such as a personal computer). The client will receive the data stream and (after a short buffering delay) decode the audio and play it to a user.

Media streaming uses protocols to control the flow of information and the devices that provide (stream) the media. Examples of media streaming protocols include real time protocol (timing information), real time control protocol (QoS information) and real time streaming protocol (media server flow control)

## Stored Media Distribution

Stored media distribution is the transfer of information (media) on a device or material that can be used to store and retrieve the media. Stored media includes magnetic tapes, magnetic disks, optical disks (CDROM or DVD), or stored flash memory modules.

Stored media may be in the form of read only media or recordable media. A majority of recording devices are designed to look for copyright control information (CCI) that is contained within the media programs. When a user records media from broadcast systems (such as recording a television program), the recording devices typically encodes (encrypts) the media so additional copies cannot be made from the recorded media.

# Media Distribution

Media distribution is the process of transferring information between a content providers and content users. The types of media distribution include direct distribution, multilevel (superdistribution) and peer to peer distribution.

## Direct Distribution

Direct distribution is the process of transferring information between the content owner or controller and end users without the user of intermediary distributors. An example of direct distribution is the use of a fixed-satellite service to relay programs from one or more points of origin directly to terrestrial broadcast stations without any intermediate distribution steps.

## Superdistribution

Superdistribution is a distribution process that occurs over multiple levels. Superdistribution typically involves transferring content or objects multiple times. An example of superdistribution is the providing of a music file to radio stations and the retransmission of the music on radio channels.

## Peer to Peer Distribution

Peer to peer distribution is the process of directly transferring information, services or products between users or devices that operate on the same hierarchical level (usage types that have similar characteristics).

## Media Portability

Portable media is information content (e.g. digital audio) that can be transferred to one or more media players.

# Key Management

Key management is the creation, storage, delivery/transfer and use of unique information (keys) by recipients or holders of information (data or media) to allow the information to be converted (modified) into a usable form.

Key assignment is the process of creating and storing a key so a user or device is able to decode content. Key assignment may be performed by sending a command from a system to a device or it may be performed by providing information to a user for direct entry into a device or system that creates a new unique key.

Key revocation is the process of deleting or modifying a key so a user or device is no longer able to decode content. Key revocation may be performed by sending a command from a system to a device.

DRM systems can have multiple types of keys with different levels (hierarchy) of security. A key hierarchy is a structure of key levels that assigns higher priority or security level to keys at higher levels. Upper layer keys may be used to encrypt lower level keys for distribution. Examples of keys with different levels include session keys that do not change during a communication session and content keys that repeatedly change every few seconds to ensure that media will stop playing if the authorization to view has ended.

Key exchange protocols are used to coordinate and securely transfer keys in a communication system. Key protocols may be proprietary or they may be standardized. An example of an industry standard key exchange protocol is Internet key exchange (IKE). IKE is defined in request for comments 2409 (RFC 2409).

## Smart Card

A smart card is a portable credit card size device that can store and process information that is unique to the owner or manager of the smart card. When the card is inserted into a smart card socket, electrical pads on the card connect it to transfer Information between the electronic device and the card. Smart cards are used with devices such as mobile phones, television set top boxes or bank card machines. Smart cards can be used to identify and validate the user or a service. They can also be used as storage devices to hold media such as messages and pictures.

## Virtual Card

To overcome the requirement of providing a physical smart card to a user, virtual smart cards may be created. A virtual smart card is a software program. A virtual smart card associates secret information on a users device (such as a TV set top box) that can store and process information from another device (a host) to provide access control and decrypt/encrypt information that is sent to and/or from the device.

# Digital Rights Management Threats

Some of the potential threats to digital rights management systems include file ripping, hacking, spoofing, hijacking and bit torrents.

Piracy (1-transmission) The operation of unauthorized commercial stations, usually in international waters near target regions. Pirates violate flu frequency regulations, copyright laws, and national broadcasting laws. (2-programming) The duplication and sale of program materials, particularly TV programs on videotape, in violation of copyright laws. (3-signals) The unauthorized use of cable TV or satellite signals.

## Ripping

Ripping media is the process of extracting (ripping it from its source) or storing media as it is streamed. Ripping media may occur through the use of a stream recorder. Stream recorders are devices and/or software that is used to capture, format and store streaming media.

Figure 9.21 shows how users may rip media from its original packaging to alter its form and potentially change or eliminate rights management attributes. This example shows that an audio CD is inserted into a personal computer that has had ripping software installed. The ripping software instructs the microprocessor in the computer to make a copy of the data that is being played to the sound card (audio chip) in the computer and to store this data on the storage disk in the computer.

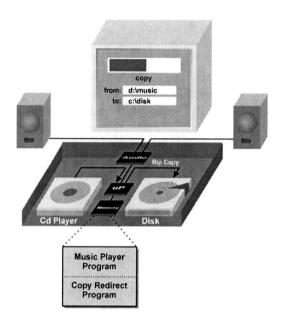

Figure 9.21, Ripping Media

## Hacking

Hacking is the process that is used when attempt to gain unauthorized access into networks and/or computing devices. The term hacking has also been used by programmers to solve their programming problems. They would continually change or hack the program until it operated the way that they desire it to operate.

Hackers perform hacking. A hacker is someone or a machine that attempts to gain access into networks and/or computing devices. Hackers may perform their actions for enjoyment (satisfaction), malicious reasons (revenge), or to obtain a profitable gain (theft).

## Spoofing

Spoofing is the use of another person's name or address to hide their true identity. Spoofing may involve registering using one name or identify and obtaining access to the media through the use of other names.

## Hijacking

Hijacking is a process of gaining security access by the capture of a communication link in mid-session after the session has already been unauthorized. Hijacking occurs when an unauthorized user detects (sniffs) and obtains information about a communication system.

A sniffer is a device or program that receives and analyzes communication activity so that it can display the information to a person or communication system. While sniffers may be used for the analysis of communication systems, they are often associated with the capturing and displaying of information to unauthorized recipients. For example, a sniffer may be able to be setup to look for the first part of any remote login session that includes the user name, password, and host name of a person logging in to another machine. Once this information is captured and viewed by the unauthorized recipient (an intruder or hacker), he or she can log on to that system at will.

Figure 9.22 shows how hijacking may be used to obtain access to an authorized media session to gain access to protected media. This example shows that an unauthorized user has obtained information about a media session request between a media provider (such as an online music store) and a user (music listener). After the media begins streaming to the validated user, the hijacker modifies a routing table distribution system to redirect the media streaming session to a different computer.

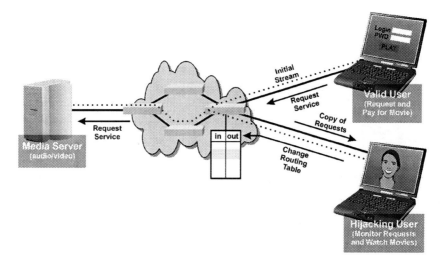

Figure 9.22, Hijacking Operation

## Bit Torrent

A bit torrent is a rapid file transfer that occurs when multiple providers of information can combine their data transfer into a single stream (a torrent) of file information to the receiving computer.

Figure 9.23 shows how to transfer files using the torrent process. This example shows that 4 computers contain a large information file (such as a movie DVD). Each of the computers is connected to the Internet via high-speed connections that have high-speed download capability and medium-speed upload capability. To speed up the transfer speed for the file transfer, the receiver of information can request sections of the media file to be downloaded. Because the receiver of the information has a high-speed download connection, the limited uplink data rates of the section suppliers are combined. This allows the receiver of the information to transfer the entire file much faster.

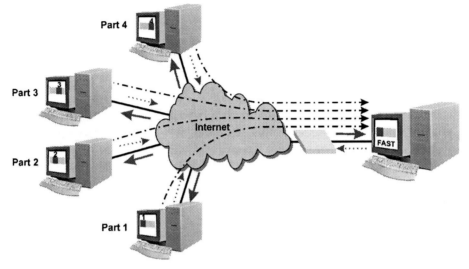

Figure 9.23, Torrent Operation

## Camming

Camming is the use of a video-recording device to capture audiovisual content from a display (e.g. using a camcorder to record a move).

## Insider Piracy

Insider piracy is the duplication and sale of program materials, particularly TV programs on DVD, videotape or other digital media format in violation of copyright laws which is performed or assisted by people who are employees, contractors or supporting personnel of companies that produce, distribute or manage the program materials.

## Analog Hole

The analog hole is the ability for users to copy media in its analog form after it has been converted from digital format to an analog signal. An example of the analog hole is the recording of an audio disk by using a microphone, which is placed next to the speakers of a digital stereo system.

### Digital Hole

The digital hole is the ability for users to redirect media that is in its digital form after it has been received and decoded and reformatted to a display device (such as a HDMI interface). An example of the digital hole is the redirecting of a digital display and processing it into a recorded format.

### Misrouting

Misrouting is the redirection of transmission paths or routes. Misrouting may be accidental (error in a router table update) or it may be intentional through the altering or changing of the routing tables.

## Protocols and Industry Standards

Industry standards are operational descriptions, procedures or tests that are part of an industry standard document or series of documents that is recognized by people or companies as having validity or acceptance in a particular industry. Industry standard defines protocols, file formats and other information that allows systems and products to interact reliably with each other. Industry standards are commonly created through the participation of multiple companies that are part of a professional association, government agency or private group.

For digital rights management in IPTV systems, digital rights standards can be used defines how the rights of intellectual property will be transferred and enforced as media is transferred from the content owner to the user. These rights and the rules associated with them are expressed using different protocols. Rights are defined in a rights expression language (REL) and rights languages can use data dictionaries to ensure that the commands or processes are defined in a common way.

Rights expression language (REL) are the processes and procedures that are used to define the rights to content, fees or other consideration required to secure those rights, types of users qualified to obtain those rights, and other associated information necessary to enable e-commerce transactions in content rights.

A rights data dictionary is a set of terms that are used by a rights expression language (REL). Using a standard RDD helps to ensure a rights management system will perform processes with an expected outcome (less misinterpretations).

Digital rights management industry standards include extensible rights management language (XrML), extensible media commerce language (XMCL), open digital rights language (ODRL), MPEG-7, MPEG-21, contracts expression language (CEL), secure digital music initiative (SDMI), and resource description framework (RDF)

## Extensible Rights Management Language (XrML)

Extensible rights management language is a XML that is used to define rights elements of digital media and services. Content Guard initially developed extensible Rights Markup Language. Several companies including Microsoft have endorsed its use. The XrML language provides a universal language and process for defining and controlling the rights associated with many types of content and services.

XrML started as digital property rights language (DPRL). It was invented by Mark Stefik of Xerox's Palo Alto research center in the mid 1990s and has transformed into extensible rights markup language (XrML). The initial version of DPRL was LISP based and 2nd version (2.0) was based on XML to allow the flexibility of describing new forms of data and processes.

Because XrML is based on extensible markup language (XML), XrML files can be customized for specific applications such as to describe books (ONIX) or web based media (RDF). For more information on XrML see www.XrML.org.

## Extensible Media Commerce Language (XMCL)

Extensible media commerce language is a software standard that is used to define exchangeable elements multimedia content and digital rights associated with the media so the media can be exchanged (interchanged) with networks or business systems.

## Open Digital Rights Language (ODRL)

Open digital rights language is an XML industry specification that is used to manage digital assets and define the rights associated with those digital assets. Open digital rights language was initially designed by Renato Iannella of IPR Systems Ltd. of Australia and its use has been endorsed by several companies including Nokia. The ODRL is a relatively simple language that has been optimized for the independent definition of rights associated with many types of content and services. For more information on ODRL visit www.ODRL.net.

## MPEG-7

The MPEG-7 multimedia content description interface is a multimedia standard that adds descriptive content to media objects.

## MPEG-21

MPEG-21 is a multimedia specification that adds rights management capability to MPEG systems. The MPEG-21 system is based on XrML.

## Contracts Expression Language (CEL)

Contracts expression language is a set of commands and procedures that are used to express the terms of contractual agreements in a text based extensible markup language (XML) format. Because CEL expressions are text based, they are understandable and usable by both humans and machines.

## Secure Digital Music Initiative (SDMI)

Secure digital music initiative (SDMI) is a digital rights management standards development project that is intended to control the distribution and ensure authorized use for digital music. Members of SDMA come from information technology, consumer electronics, security, recording companies, and Internet service provider industries. The focus (charter) of SDMA is to develop open industry specifications that protect the rights of media content owners and publishers for storing, playing and distributing of digital media while allowing users convenient access to media content through new emerging distribution systems (such as the Internet).

The initial development for SDMI (phase I) was to create a digital watermark system. Digital music access technology is a trademark for the technology used for the content protection process developed by the secure digital music initiative (SDMI).
For additional information on SDMI, visit www.SDMI.org.

## Resource Description Framework (RDF)

Resource description framework is a general-purpose descriptive language that is used to describe and manage content on the web. The RDF language is well suited for describing resources that are stored on the web such as title, author and date. RDF is managed by the

## World Wide Web consortium (W3C).

The World Wide Web consortium (W3C), formed in 1994, is a group of companies in the wireless market whose goal is to provide a common markup language for wireless devices and other small devices with limited memory. For more information about W3C and what they are endorsing as a web based DRM system can be found at www.W3C.com.

# Copy Protection Systems

Content Protection is the process or system that is used to prevent content from being pirated or tampered with in devices or within a communication network (such as in a television system). Content protection involves uniquely identifying content, assigning the usage rights, scrambling and encrypting the digital assets prior to play-out or storage (both in the network or end user devices) as well as the delivery of the accompanying rights to allow legal users to access the content.

The copy protection system may be used to control the output or media quality on specific outputs. These protected outputs are interface points in a device or system that are not directly accessible or usable to make copies of media.

Copy protection system can be designed for analog or digital media. Analog copy protection system may restrict or distort the video signal when media is copied on unauthorized devices. Digital copy protection systems can use authentication and encryption processes to ensure media is not copied or used in an unauthorized way.

There are many types of copy protection systems including copy control information (CCI), copy generation management system (CGMS), broadcast flag, serial copy management system (SCMS), content protection for prerecorded media (CPPM), content protection for recordable media (CPRM), content scrambling system (CSS), secure video processing (SVP), digital transmission content protection (DTCP), high bandwidth digital content protection (HDCP), high definition multimedia interface (HDMI) and extended conditional access (XCA).

## Copy Control Information (CCI)

Copy control information is the use of data that is provided within or in addition to media that is used to enable or disable the ability of devices to make copies of media. The CCI system uses two bits to define the copyrights associated with media. The copy control bits indicate if the media is authorized to be copied and unlimited amount of times, only one time or never. CCI may be embedded into the content so it can be read and used by devices that may be used to copy content (such as a DVD player/recorder).

## Copy Generation Management System (CGMS)

Copy generation management system is a copy control system that embeds (includes) in the media copy control information (CCI) that is can be used by replication devices (e.g. DVD burners) to determine if the media is allowed to be copied and if so, in what format types.

The CGMS system may be used for analog or digital outputs. The CGMS for analog output (CGMS-A) embeds (includes) in the media CCI in the vertical blanking interval of the video signal (lines 20 or 21). The CGMS for digital output (CGMS-D) embeds (includes) in the media CCI in the digital video protocols such as digital transmission copy protection (DTCP or high definition multimedia interface (HDMI).

## Broadcast Flag

A broadcast flag is a data word that is sent in a broadcast channel (such as a digital television channel) that indicates who can use the content. The broadcast flag defines if digital copying is allowed and if additional copies can be made (serialized copying).

## Serial Copy Management System (SCMS)

Serial copy management system is a copy control system that uses information contained in the media (e.g. in the disk) to determine if initial copies or second-generation copies are permitted. SCMS uses permission flags that indicate if the media is allowed to be copied never, copied once or copied all (unlimited). SCMS has been used in digital audiotape (DAT) recorders and in mini-disc recorders.

## Content Protection for Prerecorded Media (CPPM)

Content protection for prerecorded media is a copy protection system that is used to control copies of stored media (such as DVDs).

## Content Protection for Recordable Media (CPRM)

Content protection for recordable media is a copy protection system that is used to control copies of recordable media (such as blank DVDs).

## Content Scrambling System (CSS)

A content scrambling system is a method of encryption that is used for stored media (e.g. DVDs) to prevent the use or copying of the media information. To decode and play the information, a player with a decoding algorithm must be used. The CSS process is designed to prevent access to the media from devices that do not have the appropriate decoding algorithm so the media cannot be copied or manipulated by unauthorized users and devices (e.g. via DVD burners).

## Secure Video Processing (SVP)

Secure video processing is a method of ensuring the rights of video content are enforced through the use of hardware, software and rights management systems.

A secure boot loader is a process that is used when a system is started that ensures operations with a system is secure. A secure book loader system is designed to be tamper proof.

## Digital Transmission Content Protection (DTCP)

Digital transmission content protection is a copy control system that was developed by Intel, Sony, Hitachi, Matsushita and Toshiba for the IEEE 1394 (Firewire) system. Devices that use DTCP exchange authentication certificates and keys to setup a secure communication session. DTCP has 2 levels of security, full and restricted. Full authentication is used when the devices have enough processing power to perform authentication and encryption.

## High Bandwidth Digital Content Protection (HDCP)

High bandwidth digital content protection is the commands and processes that are used by the digital video interface (DVI) system to perform authentication, encryption and authorization of services. The HDCP system allows devices to identify each other and determine what type of device it is (e.g. a display or a recorder).

## High Definition Multimedia Interface (HDMI)

High definition multimedia interface is a specification for the transmission and control of uncompressed digital data and video for computers and high-speed data transmission systems. HDMI is a combination of a digital video interface (DVI) connection along with the security protocol HDCP. HDMI has the ability to determine if security processes are available on the DVI connection and if not, the HDMI interface will reduce the quality (lower resolution) of the digital video signal.

## Extended Conditional Access (XCA)

Extended conditional access is a digital transmission content protection is a copy control system that was developed by Zenith and Thomson for one-way and two-way digital interfaces. It has the capability to use and renew security on smart cards.

# Appendix 1 - Acronyms

**3DES**-Triple Data Encryption Standard
**60i**-60 Interlaced
**A/V**-Audio Visual
**AAC**-Advanced Audio Codec
**AAF**-Advanced Authoring Format
**AALn**-ATM Adaptation Layer n
**AC3**-Audio Compression 3
**Ad Credits**-Advertising Credits
**Ad Model**-Advertising Model
**Ad Splicer**-Advertising Splicer
**ADC**-Analog to Digital Converter
**ADET**-Aggregate Data Event Table
**ADPCM**-Adaptive Differential Pulse Code Modulation
**ADSL**-Asymmetric Digital Subscriber Line
**AEE**-Application Execution Environment
**AEIT**-Aggregate Event Information Table
**AES**-Advanced Encryption Standard
**AES**-Audio Engineering Society
**AFI**-Authority and Format Identifier
**AFX**-Animation Framework eXtension
**AI**-Artificial Intelligence
**AIR**-Application Information Resource
**AIT**-Application Information Table
**AJAX**-Asynchronous Javascript and XML
**A-LAN**-Appartment Local Area Network
**Analog HDTV**-Analog High Definition Television
**ANSI**-American National Standards Institute
**AoD**-Advertising on Demand
**API**-Application Program Interface
**APS**-Analog Protection System

**ARIB**-Association for Radio Industries and Business
**ASCII**-American Standard Code for Information Interchange
**ASF**-Advanced Streaming Format
**ASF**-Advanced Systems Format
**ASI**-Asynchronous Serial Interface
**ASN**-Abstract Syntax Notation
**ASP**-Advanced Simple Profile
**ATA**-Advanced Technology Attachment
**ATIS**-Alliance for Telecommunications Industry Solutions
**ATM**-Asynchronous Transfer Mode
**ATSC**-Advanced Television Systems Committee
**ATV**-Analog Television
**ATVEF**-Advanced Television Enhancement Forum
**AU**-Audio File Format
**AUX**-Auxiliary
**Avail**-Advertising Availability
**AV**-Audio Video
**AVI**-Audio Video Interleaved
**AVT**-AV Transport Service
**AWT**-Abstract Widowing Toolkit
**B2B Advertising**-Business to Business Advertising
**B2B**-Business to Business
**B2C**-Business to Consumer
**BCG**-Broadband Content Guide
**BCG**-Broadcast Cable Gateway
**BFS**-Broadcast File System
**BGP**-Border Gateway Protocol
**BIFF**-Binary Interchange File Format

**BIOP-**Basic Input Output Protocol
**BLC-**Broadband Loop Carrier
**BML-**Broadcast Markup Language
**BMP-**BitMaP
**BOF-**Business Operations Framework
**BPP-**Bits Per Pixel
**BPS-**Bits Per Second
**Broadband TV-**Broadband Television
**bslbf-**Bit Serial Leftmost Bit First
**BTA-**Broadcasting Technology Association in Japan
**BTSC-**Broadcast Television Systems Committee
**BUC-**Buffer Utilization Control
**CableCARD-**Cable Card
**CA-**Central Arbiter
**CA-**Certificate Authority
**CA-**Conditional Access
**CAT-**Conditional Access Table
**CATV-**Cable Television
**CCBS-**Customer Care And Billing System
**CC-**Closed Caption
**CCIR-**Comite' Consultatif International de Radiocommunications
**CCTV-**Closed-Circuit Television
**CDA-**Compact Disc Audio
**CDA-**Content Distribution Agreement
**CDATA-**Character Data
**CDI-**Content Digital Item
**CDR-**Common Data Representation
**CDS-**Content Directory Service
**CEA-**Consumer Electronics Association
**CE-**Consumer Electronics
**CEPCA-**Consumer Electronics Powerline Communication Alliance
**CERN-**Conseil European pour la Recherche Nucleaire
**CF-**Compact Flash
**CG-**Character Generator
**CGM-**Consumer Generated Media
**CI-**Common Interface
**CLUT-**Color Look-Up Table

**CM-**Configuration Management
**CM-**Connection Manager Service
**CMTS-**Cable Modem Termination System
**CN-**Core Network
**Coax Amp-**Coaxial Amplifier
**COFDM-**Coded Orthogonal Frequency Division Multiplexing
**COM-**Common Object Model
**Co-op-**Cooperative Advertising
**CORBA-**Common Object Request Broker Architecture
**CP-**Control Point
**CPL-**Composition Playlist
**CPU-**Central Processing Unit
**CRC-**Cyclic Redundancy Check
**CRID-**Content Reference Identifier
**CRL-**Certificate Revocation List
**CRLF-**Carriage Return Followed by a Line Feed
**CRT-**Cathode Ray Tube
**CSM-**Component Splice Mode
**CSP-**Communication Service Provider
**CSRC-**Contributing SouRCe
**CSS-**Cascading Style Sheets
**CSS-**Content Scramble System
**CSS-**Customer Support System
**CTAB-**Cable Television Advisory Board
**CTEA-**Copyright Term Extension Act
**CTL-**Certificate Trust List
**CUTV-**Catch Up Television
**CVBS-**Color Video Blank and Sync
**CVCT-**Cable Virtual Channel Table
**DAC-**Digital To Analog Converter
**DAI-**Digital Item Adaptation
**DASE-**DTV Application Software Environment
**DAU-**Data Access Unit
**DAVIC-**Digital Audio Video Council
**DCAS-**Downloadable Conditional Access System
**DCC Table-**Direct Channel Change Table
**DCC-**Directed Channel Change

**DCCP**-Datagram Congestion Control Protocol
**DCCT**-Discrete Channel Change Table
**DCD**-Document Content Description
**DCG**-Data Channel Gateway
**DCH**-Dedicated Channel
**DCI**-Digital Cinema Initiative
**DCM**-Device Control Module
**DCT**-Discrete Cosine Transform
**DDB**-Download Data Block
**DDE**-Declarative Data Essence
**DDI**-Data Driven Interaction
**DEBn**-Data Elementary Stream Buffer
**DEBSn**-Data Elementary Stream Buffer Size
**DECT**-Digital Enhanced Cordless Telephone
**DER**-Definite Encoding Rules
**DES**-Data Elementary Stream
**DES**-Data Encryption Standard
**DET**-Data Event Table
**DFT**-Discrete Fourier Tranform
**DHCP**-Dynamic Host Configuration Protocol
**DH**-Diffie Hellman
**DHN**-Digital Home Network
**DHS**-Digital Home Standard
**DHTML**-Dynamic Hypertext Markup Language
**DIB**-Device Independent Bitmap
**DiffServ**-Differentiated Services
**Digitizing Tablet**-Digitizing Pad
**DII**-Download Info Indication
**DLNA**-Digital Living Network Alliance
**DLP**-Discrete Logarithm Problem
**DMA Engine**-Direct Memory Access Engine
**DMA**-Digital Media Adapter
**DMA**-Direct Memory Access
**DMC**-Digital Media Controller
**DMD**-Digital Media Downloader

**DMIF**-DSM-CC Multimedia Integration Framework
**DMP**-Digital Media Player
**DMPr**-Digital Media Printer
**DMR**-Digital Media Renderer
**DMS**-Digital Media Server
**DOCSIS+**-Data Over Cable Service Interface Specification Plus
**DOCSIS®**-Data Over Cable Service Interface Specification
**DOM**-Document Object Model
**DPI**-Digital Program Insertion
**DPX**-Digital Picture eXchange
**DRAM**-Dynamic Random Access Memory
**DRM**-Digital Rights Management
**Drop Amp**-Drop Amplifier
**DSA**-Digital Signature Algorithm
**DSCP**-Differentiated Services Code Point
**DS**-Distribution Service
**DSI**-Data Service Initiate
**DSM-CC**-Digital Storage Media Command and Control
**DSM-CC-OC**-Digital Storage Media-Command and Control Object Carousel
**DSM-CC-UU**-Digital Storage Media-Command and Control User to User
**DSM**-Digital Storage Media
**DSNG**-Digital Satellite News Gathering
**DSP**-Digital Signal Processor
**DSS**-Digital Satellite System
**DSS**-Digital Signature Standard
**DST**-Data Service Table
**DTC**-Direct to Consumer
**DTD**-Document Type Definition
**DTH**-Direct To Home
**DTLA**-Digital Transmission Licensing Administrator
**DTS**-Decode Time Stamp
**DTS**-Digital Theater Sound
**DTS**-Digital Theater Systems
**DTT**-Digital Terrestrial Television

DTV-Digital Television
Dub-Dubbing
DV Camcorder-Digital Video Camcorder
DV25-Digital Video 25
DVB-ASI-Digital Video Broadcast-Asynchronous Serial Interface
DVB-Digital Video Broadcast
DVB-MHP-Digital Video Broadcasting Multimedia Home Platform
DVBSI-Digital Video Broadcast Service Information
DVD-Digital Video Disc
DVE-Digital Video Effect
DVI-Digital Visual Interface
DVR-Digital Video Recorder
DVS-Digital Video Service
DWDM-Dense Wave Division Multiplexing
EAS-Emergency Alert System
EBS-Emergency Broadcast System
EBU-European Broadcasting Union
ECMA-European Commerce Applications Script
ECMA-European Computer Manufactures Association
ECM-Enterprise Content Management
ECM-Entitlement Control Messages
EDCA-Enhanced Distributed Channel Access
EDL-Edit Decision List
EDS-Extended Data Services
EDTV-Enhanced Definition Television
EE-Execution Engine
EEPROM-Electrically Erasable Programmable Read Only Memory
EFF-Electronic Frontier Foundation
EFS-Error Free Seconds
EFTA-European Free Trade Association
EIT-Event Information Table
EKE-Encrypted Key Exchange
EOD-Everything on Demand
EPF-Electronic Picture Frame

EPG-Electronic Programming Guide
ESCR-Elementary Stream Clock Reference
ES-Elementary Stream
E-Tailers-Electronic Retailers
ETM-Extended Text Message
ETSI-European Telecommunications Standards Institute
ETT-Extended Text Table
EUMID-Extended Unique Material Identifier
Euro-DOCSIS-European Data Over Cable Service Interface Specification
E-Wallet-Electronic Wallet
FAB-Fulfillment, Assurance, and Billing
FCC-Federal Communications Commission
FDC-Forward Data Channel
FDDI-Fiber Distributed Data Interface
FES-Front End Server
FIPS-Federal Information Processing Standards
FMC-Fixed Mobile Convergence
FOD-Free on Demand
Forward OOB-Forward Out of Band Channel
FOSS-Free Open Source Software
FourCC-Four Character Code
FPA-Front Panel Assembly
FTA-Free to Air
FTP-File Transfer Protocol
FTTC-Fiber To The Curb
FW-Firmware
GCT-Global Color Table
GIF-Graphics Interchange Format
GP-Graphics Processor
GPS-Global Positioning System
GSM-Global System For Mobile Communications
GUI-Graphic User Interface
GXF-General eXchange Format
HAL-Hardware Abstraction Layer

**HANA**-High-Definition Audio-Video Network Alliance
**HAVi**-Home Audio Video Interoperability
**HCCA**-HCF Coordination Channel Access
**HCCA**-Hybrid Coordination Function Controlled Channel Access
**HCNA**-HPNA Coax Network Adapter
**HD**-High Definition
**HD-PLC**-High Definition Power Line Communication
**HDTV**-High Definition Television
**HFC**-Hybrid Fiber Coax
**HID**-Home Infrastructure Device
**HITS**-Headend in the Sky
**HMS**-Headend Management System
**HMS**-Hosted Media Server
**HND**-Home Network Device
**HomePlug AV**-HomePlug Audio Visual
**HomePNA**-Home Phoneline Networking Alliance
**HP**-High Profile
**HSM**-Hierarchical Storage Management
**HSTB**-Hybrid Set Top Box
**HTML**-Hypertext Markup Language
**HTTP**-Hypertext Transfer Protocol
**HTTPS**-Hypertext Transfer Protocol Secure
**HVN**-Home Video Network
**iAD**-Interactive Advertisements
**IANA**-Internet Assigned Numbering Authority
**ICAP**-Interactive Communicating Application Protocol
**ICC**-International Color Consortium
**ICG**-Interactive Cable Gateway
**IC**-Integrated Circuit
**IDCT**-Inverse Discrete Cosine Transform
**IDEA**-International Data Encryptions Algorithm
**IDE**-Integrated Development Environment
**IDE**-Integrated Drive Electronics

**IDL**-Interface Definition Language
**IEEE**-Institute Of Electrical And Electronics Engineers
**IETF**-Internet Engineering Task Force
**IF Switching**-Intermediate Frequency Switching
**IF**-Intermediate Frequency
**IGMP**-Internet Group Management Protocol
**IHDN**-In-Home Digital Networks
**IIC**-Inter-Integrated Circuit Bus
**IIF**-IPTV Interoperability Forum
**IIOP**-Internet Inter-ORB Protocol
**IKE**-Internet Key Exchange
**IMS**-IP Multimedia System
**Inline Amp**-Inline Amplifier
**InstanceID**-Instance Identifier
**IOR**-Interoperable Object Reference
**IP STB**-Internet Protocol Set Top Box
**IPCATV**-Internet Protocol Cable Television
**IPDC**-Internet Protocol Datacasting
**IPDC**-Internet Protocol Device Control
**IPG**-Interactive Programming Guide
**IP**-Internet Protocol
**IPMP**-Intellectual Property Management and Protection
**IPPV**-Impulse Pay Per View
**IPR**-Intellectual Property Rights
**IPTV**-Internet Protocol Television
**IPTV**-IP Television Service
**IPVBI**-IP Multicast over VBI
**IR Blaster**-Infrared Blaster
**IR Receiver**-Infrared Receiver
**IRD**-Integrated Receiver and Decoder
**IRT**-Integrated Receiver and Transcoder
**ISAN**-International Standard Audiovisual Number
**ISBN**-International Standard Book Number
**ISDB**-Integrated Services Digital Broadcasting

ISDN-Integrated Services Digital Network
IS-Intensity Stereo
ISMA-Internet Media Streaming Alliance
ISO-International Standards Organization
ISP-Internet Service Provider
ISR-Interrupt Service Routine
ITU-International Telecommunication Union
iTV-Internet TV
IVG-Integrated Video Gateway
IWS-Initial Working Set
JAAS-Java Authencation and Authroization Service
JAR-Java Archive
JCA-Java Cryptography Architecture
JCE-Java Cryptography Extentions
JCIC-Joint Committee on Intersociety Coordination
JDK-Java Development Kit
JFIF-JPEG File Interchange Format
JMF-Java Media Framework
JNG-JPEG Network Graphics
JNI-Java Native Interface
JNM-Java Native Methods
JPEG2000-Joint Picture Experts Group 2000
JPEG-Joint Photographic Experts Group
Jscript-JavaScript
JS-Joint Stereo
JSSE-Java Secure Socket Extension
JVM-Java Virtual Machine
kbps-Kilo bits per second
KDF-Key Derivation Function
Killer App-Killer Application
KLV-Key Length Value
KoD-Karaoke on Demand
KPI-Key Performance Indicator
KQI-Key Quality Indicators
LAES-Lawfully Authorized Electronic Surveillance
LAN-Local Area Network
LBI-Late Binding Interface

LCD-Liquid Crystal Display
LCN-Logical Channel Number
LFE-Low Frequency Enhancement
LID-Local Identifier
Linear TV-Linear Television
Liquid LSP-Liquid Label Switched Path
LISP-LIS Processing
LLC-SNAP-Logical Link Control-Sub Network Access Protocol
LLU-Local Loop Unbundling
LMDS-Local Multichannel Distribution Service
LMDS-Local Multipoint Distribution System
LOC-Local Operations Center
LP-1-Local Primary Monitoring Station First
LP-2-Local Primary Monitoring Station Alternate
LSD-Logical Screen Descriptor
LSF-Low Sampling Frequency
MAC-Media Access Control
MAC-Medium Access Control
MAC-Message Authentcation Code
MAP-Media Access Plan
MBGP-Multicast Border Gateway Protocol
MB-Media Block
MBONE-Multicast Backbone
Mbps-Millions of bits per second
MCard-Multiple Stream CableCARD
M-CMTS-Modular Cable Modem Termination System
MCNS-Multimedia Cable Network System
MCR-Master Control Room
MDCT-Modified Discrete Cosine Transformation
MDD-Metadata Dictionary
MDI-Media Delivery Index
MER-Modulation Error Ratio
MGG-LC-Low Complexity MNG
MG-Minimum Guarantee

**MGT**-Master Guide Table
**MHD**-Mobile Handheld Device
**MHEG**-Multimedia/Hypermedia Expert Group
**MHP**-Multimedia Home Platform
**Microchannel**-Television Micro Channel
**MIDI**-Musical Instrument Digital Interface
**MIME**-Multipurpose Internet Mail Extensions
**MITRE**-Missile Test and Readiness Equipment
**MIU**-Media Interoperability Unit
**MJPEG**-Motion JPEG
**MMDS**-Multichannel Multipoint Distribution Service
**MMDS**-Multipoint Microwave Distribution System
**MMS**-Microsoft Media Server Protocol
**MMU**-Memory Management Unit
**MNG**-Multiple image Network Graphics
**MoCA**-Multimedia over Coax Alliance
**MOV**-QuickTime MOVie format
**MP3**-Motion Picture Experts Group Layer 3
**MP3**-Motion Picture Experts Group Level 3
**MP4**-MPEG-4
**MPEG**-Motion Picture Experts Group
**MPEGoIP**-MPEG over Internet Protocol
**MP**-Main Profile
**MPTS Feed**-Multiprogram Transport Stream Feed
**MPTS**-Multiprogram Transport Stream
**MRD**-Marketing Requirements Document
**MRD**-MPEG-2 Registration Descriptor
**MRLE**-Microsoft Run Length Encoding
**MSB**-Most Significant Bit
**MSDP**-Multicast Source Discovery Protocol
**MSE**-Mean Square Error
**MS**-Media Server
**MSO**-Multiple System Operator

**MTFTP**-Multicast Trivial File Transfer Protocol
**MTT**-Mobile Terrestrial Television
**MTU**-Maximum Transmission Unit
**MuX**-Multiplexer
**MVDDS**-Multichannel Video Distribution and Data Service
**MVPD**-Multichannel Video Program Distributor
**MXF**-Material eXchange Format
**NAB**-National Association Of Broadcasters
**NABTS**-North American Basic Teletext Specification (EIA_516)
**NAL**-Network Abstraction Layer
**NAN**-Not A Number
**NCF**-Network Connectivity Function
**NCTA**-National Cable Television Association
**Net**-Internet
**NICAM**-Near Instantaneous Companded Audio Multiplexing
**NIC**-Network Interface Card
**NI**-Network Interface
**NISDN**-Narrrow-band Integration Services Digital Network
**NIST**-National Institute Of Standards And Technology
**NNW**-No New Wires
**NOC**-National Operations Center
**NOC**-Network Operations Center
**NPT**-Normal Play Time
**NPVR**-Network Personal Video Recorder
**NRT**-Network Resources Table
**NSAP**-Network Service Access Point
**NTP**-Network Time Protocol
**NTSC**-National Television System Committee
**NUT**-Net UDP Throughput
**NVOD**-Near Video On Demand
**OBE**-Out of Box Experience
**OCAP**-Open Cable Application Platform

**OC**-Object Carousel
**On Airwaves**-On-Air
**OOB Channel**-Out of Band Channel
**OOB Receiver**-Out of Band Receiver
**OPERA**-Open PLC European Research Alliance
**ORB**-Object Request Broker
**OSD**-On Screen Display
**OSGI**-Open Systems Gateway Initiative
**OS**-Operating System
**OSPF**-Open Shortest Path First
**OSS**-Operations Support System
**OUI**-Organization Unique Identifier
**P2P**-Peer to Peer
**PAL**-Phase Alternating Line
**PAM**-Pulse Amplitude Modulation
**Parametric QoS**-Parametric Quality of Serivce
**PAT**-Program Association Table
**PBS**-Public Broadcast Service
**PCDATA**-Parsed Character Data
**PCI**-Peripheral Component Interconnect
**PCMCIA**-Personal Computer Memory Card International Association
**PCM**-Pulse Coded Modulation
**PCR**-Production Control Room
**PCR**-Program Clock Reference
**PDA**-Personal Digital Assistant
**PDM**-Pulse Duration Modulation
**PDV**-Packet Delay Variation
**PEM**-Privacy Enhanced Mail
**PE**-Presentation Engine
**PER**-Packet Error Rate
**PES**-Packetized Elementary Stream
**PFR**-Portable Font Resource
**PGP**-Pretty Good Privacy
**PID Dropping**-Packet Identifier Dropping
**PID**-Packet Identifier
**PIG**-Picture in Graphics
**PIMDM**-Protocol Independent Multicase Dense Mode
**PIM**-Protocol Independent Multicast

**PIP**-Picture in Picture
**PKI**-Public Key Infrastructure
**PKIX**-PKI X.509
**Play-List**-Playlist
**PLC**-Power Level Control
**PLMN**-Public Land Mobile Network
**PLTV**-Pause Live Television
**PMI**-Portable Media Interface
**PMP**-Portable Media Player
**PMS**-Personal Media Server
**PMS**-Property Management System
**PMT**-Program Map Table
**PNG**-Portable Network Graphics
**POD**-Personal Operable Device
**POD**-Point of Deployment
**POD**-Point of Deployment Module
**POE**-Point of Entry
**POTS**-Plain Old Telephone Service
**PPD**-Pay Per Day
**PPV**-Pay Per View
**Preamp**-Pre-Amplifier
**Private TV**-Private Television
**PSIP**-Program and System Information Protocol
**PSI**-Program Specific Information
**PSM**-Program Splice Mode
**PSNR**-Peak Signal to Noise Ratio
**PSTD**-Program System Target Decoder
**PSTN**-Public Switched Telephone Network
**PSU**-Pillow Speaker Unit
**PTS**-Presentation Time Stamp
**PU**-Presentation Unit
**Push VOD**-Push Video on Demand
**PVR**-Personal Video Recorder
**QAM**-Quadrature Amplitude Modulation
**QoS Policy**-Quality of Service Policy
**QoS**-Quality Of Service
**QPSK**-Quadrature Phase Shift Keying
**QT Atoms**-Quicktime Atoms
**RADIUS**-Remote Access Dial In User Service

**RAI-**Resource Adaptation Engine
**RAM-**Random Access Memory
**RA-**Registration Authority
**RAS-**Remote Access Server
**RAS-**Rights Access System
**RCC-**Reverse Control Channel
**RC-**Remote Control
**RCS-**Rendering Control Service
**RDC-**Reverse Data Channel
**RDF-**Resource Description Framework
**RDP-**Remote Display Protocol
**Regional VHO-**Regional Video Hub Office
**Reverse OOB-**Reverse Out of Band Channel
**RF Bypass-**Radio Frequency Bypass Switch
**RF Modulator-**Radio Frequency Modulator
**RF Out-**Radio Frequency Output
**RFC-**Request For Comments
**RGBA-**Red Green Blue Alpha
**RGB-**Red, Green, Blue,
**RG-**Residential Gateway
**RHVO-**Remote Video Hub Operation
**RIFF-**Resource Interchange File Format
**RM-**RealMedia
**ROM-**Read Only Memory
**RPC-**Regional Protection Control
**RPF-**Reverse Path Forwarding
**RP-**Rendezvous Point
**RPTV-**Rear Projection Television
**RRT-**Rating Region Table
**RSA-**Rivest, Shamir, Adleman
**RSS-**Really Simple Syndication
**RSVP-**Resource Reservation Protocol
**RTCP-**Real-Time Transport Control Protocol
**RTOS-**Real Time Operating System
**RTP-**Real Time Protocol
**RTSP-**Real Time Streaming Protocol

**RTSPT-**Real Time Streaming Protocol over TCP
**RTSPU-**Real Time Streaming Protocol over UDP
**S/PDIF-**Sony Philips Digital InterFace
**SAN-**Storage Area Network
**SAP-**Secondary Audio Program
**SAP-**Session Announcement Protocol
**SAS-**Subscriber Authorization System
**SBB-**Set Back Box
**SBNS-**Satellite and Broadcast Network System
**SCard-**Single Stream CableCARD
**SCP-**Service Control Protocol
**SCR-**System Clock Reference
**SCSI-**Small Computer Systems Interface
**SCTE-**Society of Cable Telecommunication Engineers
**SDD-**Self Describing Device
**SDF-**Service Description Framework
**SDH-**Synchronous Digital Hierarchy
**SDI-**Serial Digital Interface
**SDK-**Software Development Kit
**SDP-**Session Description Protocol
**SDS-**Service Discovery and Selection
**SD-**Standard Definition
**SDT-**Service Description Table
**SDTV-**Standard Definition Television
**SDV-**Switched Digital Video
**SECAM-**Sequential Couleur Avec MeMoire
**SELT-**Single Ended Line Test
**SGML-**Standard Generalized Markup Language
**SGW-**Service Gateway
**SHE-**Super Headend
**SID-**Service Identifier
**SII-**Station Identification Information
**SI-**Service Information
**SI-**System Information
**SMATV-**Satellite Master Antenna Television

**SMBUS**-System Management Bus
**SMIL**-Synchronized Multimedia Integration Language
**SMS**-Screen Management System
**SMS**-Subscriber Management System
**SNMP**-Simple Network Management Protocol
**SOC**-System On Chip
**Softkeys**-Soft Keys
**SPG**-Synchronization Pulse Generator
**SP**-Simple Profile
**SRM**-Session and Resource Manager
**SRTP**-Secure Real Time Protocol
**SRTS**-Synchronous Residual Time Stamp
**SSH**-Secure Shell
**STC**-System Time Clock
**Stereo**-Stereophonic
**STP**-Service Transport Protocol
**STS**-System Time Stamp
**Studio**-Movie Studio
**STV**-Subscription Television
**Stylesheet**-Style Sheet
**S-Video**-Separate Video
**SVOD**-Subscription Video on Demand
**Sync Impairment**-Synchronization Impairments
**TFS**-Transient File System
**TFS**-Transport File System
**TFTP**-Trivial File Transfer Protocol
**Timestamp**-Time Stamp
**T-Mail**-Television Mail
**TMS**-Theater Management System
**Transport ID**-Transport Identifier
**TRT**-Total Running Time
**TSFS**-Transport Stream File System
**TSTV**-Time Shift Television
**TSTV**-Time Shifted Television
**TTS**-Text To Speech
**TV Centric**-Television Centric
**TV Channel**-Television Channel
**TV Portal**-Television Portal
**TV Studio**-Television Studio

**TVoF**-Television over Fiber
**TXOP**-Transmission Opportunity
**UGC**-User Generated Content
**UI**-User Interface
**UMB**-Ultra Mobile Broadband
**UMID**-Unique Material Identifier
**UPA**-Universal Powerline Association
**Upfronts**-Upfront Advertising
**UPnP AV**-Universal Plug and Play Audio Visual
**UPnP**-Universal Plug and Play
**USC**-User Selectable Content
**USM**-User Services Management
**VBI**-Video Blanking Interval
**vBook**-Video Book
**VBR Feed**-Variable Bit Rate Feed
**VBV**-Video Buffer Verifier
**VCL**-Video Coding Layer
**VCO**-Video Central Office
**VCT**-Virtual Channel Table
**VC**-Virtual Channel
**VDN**-Video Distribution Network
**V-Factor**-V Factor
**VHO**-Video Hub Office
**Video Ringtone**-Video Ring Tone
**Virtual TV Channel**-Virtual Television Channel
**VLE**-Variable Length Encoding
**VM**-Virtual Machine
**VOD**-Video On Demand
**VQI**-Video Quality Index
**VQM**-Video Quality Measurement
**VRML**-Virtual Reality Modeling Language
**VSO**-Video Serving Office
**WMA**-Windows Media Audio
**WMM**-Wi-Fi Multimedia
**WM**-Windows Media
**WTCG**-Watch This Channel Grow
**X3D**-Extensible 3D
**XDI**-Context Digital Item
**XSLT**-XML Stylesheet Language Transformation

# Appendix 2 - Standards Groups

## IPTV Standards Groups

A standards group is organization or firm that is charted to create or assist in the creation of product or service operational, performance and other types of specifications. Standards groups usually represent the interests of a group of companies that want to create products or services that are interoperable with each other.

## Motion Picture Experts Group (MPEG)

Motion picture experts group (MPEG) standards are digital video encoding processes that coordinate the transmission of multiple forms of media (multimedia). Motion picture experts group (MPEG) is a working committee that defines and develops industry standards for digital video systems. These standards specify the data compression and decompression processes and how they are delivered on digital broadcast systems. MPEG is part of International Standards Organization (ISO).

## Society of Motion Picture and Television Engineers (SMPTE)

The SMPTE is an organization that assists with the development of standards for the motion picture and television industries.

## European Telecommunications Standards Institute (ETSI)

An organization that assists with the standards-making process in Europe. They work with other international standards bodies, including the International Standards Organization (ISO), in coordinating like activities.

## Advanced Television Systems Committee (ATSC)

The Advanced Television Systems Committee (ATSC) is an international non-profit organization that assists with the development of advanced television technologies and systems. The membership of the ATSC represents a wide range of broadcast and consumer electronics segments including broadcasting, motion pictures, computers, consumer electronics, and satellite. ATSC was founded in 1982.

ATSC industry standard include digital television (DTV), high definition television (HDTV), standard definition television (SDTV), data broadcasting, surround sound audio, and satellite broadcasting.

## Association for Radio Industries and Business (ARIB)

Association for Radio Industries and Business (ARIB) is an association in Japan that oversees the creation of telecommunications standards.

## Digital Video Broadcast (DVB)

Digital video broadcasting is the sending of television signals over digital transmission channels. DVB transmission can be over different types of systems including broadcast radio, satellite systems, cable television systems and mobile communications. DVB industry standards that are published by the joint technical committee (JTC) of the European Telecommunications Standards Institute (ETSI). These standards can be obtained at www.ETSI.org.

## Cable Laboratories, Inc. (CableLabs)

Cablelabs® is a non profit consortium of members from the cable television industry founded in 1988 that oversees and assists in the development of technologies used in cable television systems.

## Advanced Television Enhancement Forum (ATVEF)

Advanced Television Enhancement Forum (ATVEF) is a group of companies and industry expert that come from multiple industries (cross-industry) that work towards the creation of industry standards for combining Internet content and broadcast television.

## World Wide Web Consortium (W3C)

The world wide web consortium (W3C), formed in 1994, is a group of companies in the wireless market whose goal is to provide a common markup language for wireless devices and other small devices with limited memory.

## Internet Media Streaming Alliance (ISMA)

Internet media streaming alliance is an association that works to help develop and promote standards that enable streaming media applications.

## Universal Plug and Play (UPnP)

Universal plug and play is an industry standard that simplifies the installation, setup, operation removal of consumer electronic devices. UPnP includes the automatic recognition of device type, communication capability, service capabilities, activation and deactivation of software drivers and system management functions. More information about UPnP can be found at www.UPnP.org.

## Digital Living Network Alliance (DLNA)

Digital living network alliance is a group of companies that work to create a set of interoperability guidelines for digital media devices that operate in the home. More information about DLNA can be found at www.DLNA.org.

# Index

802.11e, 268-269, 297

Ad Server, 200

Ad Telescoping, 39, 49-50

Adaptive Differential Pulse Code Modulation (ADPCM), 128

Addressable Advertising, 10, 34, 39, 47-48, 200

Advanced Audio Codec (AAC), 18, 111, 126-128, 138, 164, 171, 280

Advanced Authoring Format (AAF), 320

Advanced Encryption Standard (AES), 251, 280, 334-335

Advanced Simple Profile (ASP), 172

Advanced Streaming Format (ASF), 125

Advanced Systems Format (ASF), 125

Advertiser, 48, 52-53

Advertising Availability (Avail), 200

Advertising Splicer (Ad Splicer), 199, 201

Alpha Blending, 174, 287-288

American Standard Code for Information Interchange (ASCII), 170

Analog Cue Tone, 200

Analog Television (ATV), 17, 62, 67, 71, 155, 193, 198, 207, 210, 243

Analog to Digital Converter (ADC), 112, 316

Animation Framework eXtension (AFX), 166

Application Information Resource (AIR), 101, 187, 190, 192

Application Program Interface (API), 173

Artifacts, 77, 91, 93, 268

Asset Storage, 194, 197-198, 322

Asymmetric Digital Subscriber Line (ADSL), 205-206

Asynchronous Transfer Mode (ATM), 191, 202, 212

Audio Channel, 107, 126-127

Audio Compression 3 (AC3), 107

Audio Engineering Society (AES), 251, 280, 334-335

Audio File Format (AU), 128

Audio Format, 111, 125

Audio Processing, 2, 170, 280

Audio Profile, 170-171

Audio Video (AV), 18, 144, 251-254, 324, 327-328

Audio Video Interleaved (AVI), 125

Audio Visual (A/V), 221, 226, 251

Automation, 197, 202, 239, 241

Bandpass, 103-104

Base Layer, 86, 168, 171

Baseband, 296

Baseband Audio, 296

Billing System, 56, 195

Binding, 340

Bits Per Pixel (BPP), 60, 76

Blocking, 112, 243, 245

Blurring, 93

Bookings, 194

Broadband Television (Broadband TV), 141

Broadcast Flag, 360-361

Cable Modem Termination System (CMTS), 209

Cable Telephone, 179

Cable Television (CATV), 5, 42, 132, 141-142, 179, 181, 192-193, 201, 204, 207-208, 218, 226, 242-243, 262, 265, 273, 285, 311, 323

Caching, 198

Cascading Style Sheets (CSS), 362

Cataloging, 321, 323

Central Arbiter (CA), 251, 269, 338, 340-342

Certificate Authority (CA), 251, 338, 340-342

Channel Decoder, 283

Channel Selection, 39, 42-43, 181, 284-285

Chrominance, 64-65, 70-71, 139, 292

Ciphertext, 332, 334

Client Device, 348

Coding Redundancy, 79

Color Depth, 60

Comite' Consultatif International de Radiocommunications (CCIR), 69-70

Common Object Model (COM), 360

Community Content Programming, 38

Compliance, 185

Composite Signal, 73

Compression Ratio, 111-112, 133, 163, 170-173

Conditional Access (CA), 153-154, 251, 284, 328, 338, 340-342, 361, 364

Conditional Access Table (CAT), 153-154

Consumer Electronics (CE), 359

Container, 95-97, 124, 128, 165, 319-320

Content Acquisition, 194-195

Content Directory Service (CDS), 310

Content Distribution, 183, 195

Content Encoding, 199

Content Feed, 187

Content Processing, 194, 198

Content Provider, 4, 187, 203, 345

Content Rights Management, 230, 299

Content Scramble System (CSS), 362

Content Source, 187, 343

Contribution Network, 183, 185, 189

Core Network (CN), 202-204, 215

Cross Phase Coupling, 240, 251

Cue Tones, 199

Customer Support System (CSS), 362

Data Elementary Stream (DES), 248, 264, 280, 334-335

Data Encryption Standard (DES), 248, 264, 280, 334-335

Data Over Cable Service Interface Specification (DOCSIS)®, 208

Data Over Cable Service Interface Specification Plus (DOCSIS+), 208

Decode Time Stamp (DTS), 107, 149

Digital Asset, 314, 321-324

Digital Fingerprint, 337

Digital Home Standard (DHS), 260

Digital Item Adaptation (DAI), 167

Digital Program Insertion (DPI), 60, 200, 317

Digital Rights Management (DRM), 129, 134, 165-167, 279-280, 299, 308, 328, 330, 334, 336, 342-346, 350-351, 356-357, 359-360

Digital Signal Processor (DSP), 280

Digital Signature, 339

Digital Storage Media (DSM), 94-95, 160, 197

Digital Storage Media Command and Control (DSM-CC), 94-95, 160, 164

Digital Television (DTV), 1-2, 96, 131-132, 134, 158, 163-164, 179, 192-193, 207-208, 224, 243, 275, 291, 361

Digital Terrestrial Television (DTT), 141-142, 193, 201, 278, 285, 291, 298

Digital Theater Sound (DTS), 107, 149

Digital Theater Systems (DTS), 107, 149

Digital Video Broadcast (DVB), 141

Digital Video Disc (DVD), 13, 134, 165, 172-173, 181, 288, 291, 318, 321-322, 337, 348, 354-355, 361-362

Digital Visual Interface (DVI), 286, 295, 363-364

Digital Watermark, 336-337, 359

Direct Distribution, 348-349

Direct To Home (DTH), 323

Discrete Cosine Transform (DCT), 77, 156

Display Device, 356

Display Formatting, 69, 161-162

Distribution Control, 194, 201

Distribution Service (DS), 210

Dolby AC-3®, 107

Double Hop, 225

Download and Play, 275

Download Info Indication (DII), 167

Drop, 265

DSM-CC Multimedia Integration Framework (DMIF), 165

Dual Display, 278

Dynamic Host Configuration Protocol (DHCP), 281

Echo Control, 232

Edit Decision List (EDL), 322

Educational Access Channel, 188

Effects, 15, 17, 93, 101, 107, 110, 121, 136, 155, 176, 196, 244, 246, 248, 255

Electronic Programming Guide (EPG), 42, 152, 196, 288

Elementary Stream (ES), 143-144, 146, 154

Embedded, 28, 42, 185, 199, 253, 276, 279, 282, 319, 337, 339, 361

Embedded Application, 282

Emergency Alert System (EAS), 189

Enhanced Distributed Channel Access (EDCA), 269

Enhancement Layer, 86

Error Blocks, 85, 91-92

Error Retention, 85

Essence, 44, 314, 319-320

Everything on Demand (EOD), 39, 49-51

Extended Unique Material Identifier (EUMID), 325

Extensible 3D (X3D), 174-176

Fast Slide, 225

Feed, 184, 187-188, 190

Fiber Distributed Data Interface (FDDI), 268

Fiber Ring, 203

Fiber Spur, 203

Fiber To The Curb (FTTC), 216

File Transfer Protocol (FTP), 281

Flash Memory, 348

Flat Tail Content, 39

Flicker, 61, 66, 155

Foreground, 87, 174, 287

Framework, 165-166, 357, 359

Franchise, 230

Frequency Masking, 112-113

Game Port, 287, 289

Global Roaming, 210

Government Access Channel, 188

Graphic User Interface (GUI), 176

Graphics, 34, 77, 96, 161, 170, 175-177, 198-199, 277-278, 280-281, 287, 312, 317

Graphics Processing, 198-199, 280-281, 287

Graphics Processor (GP), 199

HCF Coordination Channel Access (HCCA), 269

Helicopter Feed, 184, 187

High Definition (HD), 86, 96, 133, 140, 168-169, 173, 177-178, 191, 224, 261, 278, 295, 318, 361, 363

High Definition Power Line Communication (HD-PLC), 261

High Definition Television (HDTV), 1, 62, 96, 133, 168-169, 173, 224-225, 318

High Profile (HP), 168-169, 173

Home Banking, 43

Home Coverage, 229, 240-241

Home Phoneline Networking Alliance (HomePNA), 247, 254-260

Home Shopping, 43

HomePlug 1.0, 248, 250-251, 253

HomePlug Audio Visual (HomePlug AV), 251-254

HomePlug Specification, 247, 254

HomePNA 1.0, 255-256

HomePNA 2.0, 255-256, 258

HomePNA 3.0, 256-259

HomePNA 3.1, 258-259

Hybrid Coordination Function Controlled Channel Access (HCCA), 269

Hybrid Set Top Box (HSTB), 275

Hypertext Transfer Protocol (HTTP), 91, 120, 281, 340

Image Map, 156

Image Resolution, 60, 72-73

Infrared Blaster (IR Blaster), 288

Infrared Receiver (IR Receiver), 27, 277

Ingesting Content, 195, 321

Insertion Loss, 264

Installation, 209, 226-227, 230, 232, 246, 255

Institute Of Electrical And Electronics Engineers (IEEE), 212, 266, 290, 294, 363

Integrated Circuit (IC), 280

Integrated Receiver and Decoder (IRD), 191-192

Intellectual Property Management and Protection (IPMP), 165, 167

Intellectual Property Rights (IPR), 301-305, 358

Intelligent Streaming, 89

Intensity Stereo (IS), 1, 4, 7-12, 14-16, 18-21, 23-25, 27-28, 31-57, 59-97, 99-117, 119-129, 131-140, 142, 144-180, 183-218, 220-237, 239-249, 251-258, 260-271, 273-306, 308-364

Interaction, 53, 174-175, 227, 319, 343

Interactive Applications, 175

International Standard Audiovisual Number (ISAN), 327-328

International Standard Book Number (ISBN), 324-327

International Standards Organization (ISO), 76, 78, 127, 131, 133, 327-328

International Telecommunication Union (ITU), 72-73, 133, 210, 256

Internet Audio, 117

Internet Group Management Protocol (IGMP), 281

Internet IPTV, 283

Internet Key Exchange (IKE), 350

Internet Media Streaming Alliance (ISMA), 335

Internet Protocol (IP), 1-4, 20-21, 28, 42, 59, 88-89, 91, 95, 99, 116, 120-121, 165, 179-182, 184, 190-192, 199, 201-202, 204-207, 210, 212-213, 218-219, 221-222, 227, 237, 242, 244-245, 253, 256, 261-263, 268, 273-278, 281, 283-285, 297-298, 348

Internet Protocol Set Top Box (IP STB), 28, 237, 275, 277-278, 281, 283-285, 298

Internet Service Provider (ISP), 359

Internet TV (iTV), 43

Interstitial Ad, 34

Intersystem Interference, 227

IPTV Accessories, 282

JavaScript (Jscript), 282, 315

Joint Photographic Experts Group (JPEG), 76-77, 95, 97

Kilo bits per second (kbps), 112, 117, 126-127, 138, 209-210, 223, 255, 316

Leakage, 227, 263

Linear Television (Linear TV), 196

LIS Processing (LISP), 357

Live Feed, 188

Loader, 282, 363

Local Area Network (LAN), 3, 209-212, 217, 221, 226, 236-239, 266-269, 271, 289, 296

Local Feed, 187

Local Headend, 190

Local Multichannel Distribution Service (LMDS), 3, 209-210

Local Multipoint Distribution System (LMDS), 3, 209-210

Long Tail Content, 39

Low Frequency Enhancement (LFE), 101, 106

Main Profile (MP), 168-170, 173, 177

Master Control, 257, 261

Material eXchange Format (MXF), 320

Media Access Control (MAC), 208, 212, 251, 268-269

Media Access Plan (MAP), 60, 153-154, 156, 234, 257

Media Block (MB), 334

Media Player, 17, 19-20, 27, 88-89, 97, 119, 126, 129, 180, 225, 277, 283, 347

Media Position Indicator, 225

Media Server (MS), 7, 17-18, 20-21, 23-24, 42, 89-90, 119, 193, 202, 263, 281, 347-348

Medium Access Control (MAC), 208, 212, 251, 268-269

Memory, 19, 159, 175, 177, 198, 203, 340, 348, 360

Message Authentcation Code (MAC), 208, 212, 251, 268-269

Metadata Management, 194-195

Metadata Normalization, 196

Microsoft Media Server Protocol (MMS), 281

Microwave Link, 184

Middleware, 18, 277, 282

Middleware Client, 282

Middleware Compatibility, 277, 282

Misrouting, 356

Mosquito Noise, 93

Motion JPEG (MJPEG), 95, 97

Motion Picture Experts Group Level 3 (MP3), 111, 124, 126-127, 137-138, 163, 171, 280, 316-317

Movie Studio (Studio), 168-169, 172, 177, 182, 187, 203, 308, 311-313

MPEG-4 (MP4), 86-87, 96, 133-134, 140-142, 156, 161, 164-165, 167, 170-171, 173-176, 199, 228

Multichannel Multipoint Distribution Service (MMDS), 3, 209-210

Multimedia Computer, 26-27, 174, 276-277

Multimedia over Coax Alliance (MoCA), 247, 264-266, 272

Multiplexer (MuX), 148, 151

Multipoint Microwave Distribution System (MMDS), 3, 209-210

Multiprogram Transport Stream (MPTS), 145, 147, 191-192

Music Distribution, 224

Musical Instrument Digital Interface (MIDI), 124

Must Carry Regulations, 193

National Television System Committee (NTSC), 17, 62, 65, 67-68, 71, 163, 193, 207, 275, 280, 286, 292

Natural Audio, 170-171

Near Video On Demand (NVOD), 43, 49-50

Nearline Storage, 198

Network Access, 204

Network Connection, 183, 273, 277-278, 284-285, 290, 298

Network Content, 5

Network Feeds, 182, 187

Network Gaming, 32

Network Interface (NI), 239, 273, 284

Network Operator, 34, 328

No New Wires (NNW), 227, 266, 296

Off Air Feed, 187

Offline Editing, 322

Offline Storage, 198, 322

On Demand Programming, 31

On Screen Display (OSD), 281, 287

Online Assets, 322

Online Gaming, 32

Online Storage, 198, 322

Operating System (OS), 287, 340

Packet Identifier (PID), 144-146, 153-154

Packetized Elementary Stream (PES), 144-147

Pay Per View (PPV), 5, 312

Peer to Peer (P2P), 231, 348-349

Personal Content, 37-38

Personal Video Recorder (PVR), 283

Phase Alternating Line (PAL), 17, 65, 68-69, 71, 193, 207, 275, 280, 286, 292

Phase Coupling, 240, 251

Picture in Graphics (PIG), 278

Picture in Picture (PIP), 176, 278

Pixel Depth, 60

Platform, 251, 253

Playout, 182, 194-197, 225, 323, 360

Playout System, 196

Plug-In, 283

Point of Entry (POE), 227

Portable Media Player (PMP), 126, 277

Power Level Control (PLC), 204, 214, 232, 234

Presentation Time Stamp (PTS), 144, 149

Pretty Good Privacy (PGP), 334

Primary Event, 196

Private Television (Private TV), 5, 43, 181

Production, 133, 169-170, 202, 304, 310-313, 315, 317-318, 320, 322-323

Production Control Room (PCR), 118, 160

Program Association Table (PAT), 153-154

Program Clock Reference (PCR), 118, 160

Program Distribution, 5, 26, 32, 204

Program Guide, 273-274

Program Map Table (PMT), 153-154

Program Specific Information (PSI), 151

Programming Sources, 31, 189-190

Public Access Channel, 188

Public Key Infrastructure (PKI), 280

Public Switched Telephone Network (PSTN), 213

Pulse Coded Modulation (PCM), 128

Quadrature Amplitude Modulation (QAM), 256

Quality Of Service (QoS), 8, 180, 212, 217, 221, 226, 228, 238-239, 253, 256, 258, 264, 268-269, 297, 348

QuickTime, 95-96

QuickTime MOVie format (MOV), 95-96

Radio Frequency Output (RF Out), 291

Random Access Memory (RAM), 125, 322

Read Only Memory (ROM), 141

Real Time Protocol (RTP), 227, 348

Real Time Streaming Protocol (RTSP), 94-95, 281, 348

Really Simple Syndication (RSS), 188

RealMedia (RM), 96, 125

Red, Green, Blue,(RGB), 65, 292

Redundancy, 76, 79, 82-83, 111

Reference Frame, 81

Registration Authority (RA), 125, 327

Remediation, 231

Remote Control (RC), 21, 27, 225, 277, 288, 298

Remote Diagnostics, 231

Repository, 322

Request For Comments (RFC), 95, 342, 350

Residential Gateway (RG), 225, 255, 257

Resource Description Framework (RDF), 357, 359

Return Path, 285, 292

Rights Consideration, 309

Rights Management, 134, 154, 165-167, 229-230, 279, 299-300, 306-308, 343, 345-346, 351, 356-359, 363

Rights Model, 309-313

Rivest, Shamir, Adleman (RSA), 334

Sampling Rate, 108-109, 115, 135

Satellite Television, 132

Scalable Profile, 168, 170, 172

Scan Lines, 67, 82

SCART Connector, 293-294

Scene Graph Profile, 174-175

Scheduled Programming, 31

Secondary Content, 315

Secure Microprocessor, 280

Security Protocol, 295, 341, 363

Self Discovery, 231

Separate Video (S-Video), 66, 292-294, 298

Sequential Couleur Avec MeMoire (SECAM), 65, 69, 193

Service Information (SI), 174

Serving Area, 206-207

Session Protocols, 28, 276

Show Looking, 225

Simple Network Management Protocol (SNMP), 281

Simple Profile (SP), 168, 171-172

Small Footprint, 175

Soft Client, 277

Software Module, 282

Sony Philips Digital InterFace (S/PDIF), 296

Special Effects, 196

Speech Profile, 170

Splitter Jumping, 264-265

Sponsored Content, 38

Squeeze Back, 196

Standard Definition (SD), 86, 133, 140, 177, 278, 318

Standard Definition Television (SDTV), 177

Statistical Multiplexer, 151

Stereophonic (Stereo), 100, 107, 137-138, 296, 356

Stored Media, 49-50, 96, 132, 141, 173, 182-183, 185, 187, 283, 321, 323, 346, 348, 362

Stream Thinning, 90-91

Streaming, 17-18, 20-21, 24, 86, 88-91, 94-96, 116-118, 120, 125, 128, 142-143, 160, 162, 173, 196-197, 201, 223-224, 281, 288, 323, 335, 346, 348, 351, 353

Streaming Audio, 95, 117

Streaming Video, 89, 143, 173

Studio, 168-169, 172, 177, 182, 187, 203, 308, 311-313

Subscriber Management, 340

Subscriber Management System (SMS), 340

Super Headend (SHE), 190, 353

Surround Sound, 100-101, 107, 137, 143, 164, 280, 296

Synchronized Multimedia Integration Language (SMIL), 161, 175

Syndicated Program, 37

Syndication Feeds, 188

Synthetic Audio, 170, 316
System Information (SI), 153
System Time Clock (STC), 144
System Time Stamp (STS), 149
Telepresence, 176
Telescoping, 39, 49-50
Telescoping Advertisements, 49
Television Channel (TV Channel), 4-5, 7, 18, 41, 46-47, 96, 131, 180-181, 193, 207-208, 361
Television Set, 64, 279, 282, 287, 350
Temporal Redundancy, 76, 83
Terrestrial Television, 141-142, 291
Text To Speech (TTS), 170-171
Tiling, 91, 172
Time Shifted Television (TSTV), 204
Time Stamp (Timestamp), 144, 149, 151
TMDS®, 295
Token, 268
Track, 4, 41, 44, 180
Transcoder, 193
Transcoding, 111, 198-199, 323
Transform Coding, 72-73
Trick Mode, 225, 288
Trick Play, 225, 228, 288
Trigger, 33, 124
Triple Data Encryption Standard (3DES), 335
Trivial File Transfer Protocol (TFTP), 281
Truck Feed, 187
Tuner, 28, 193, 276, 285, 291-292
Tuning Head, 193
TV RF Output, 291
Unique Material Identifier (UMID), 324-325
Universal Powerline Association

(UPA), 198, 260
Upgradability, 231, 283
Usage Rights, 194, 360
User Defined, 154
User Interface (UI), 161, 175-176, 273-274, 287
Variable Length Encoding (VLE), 79-80
Video Broadcasting, 291
Video Capture, 73
Video Channel, 21
Video Encoder, 158, 193
Video Format, 27, 69-73, 95-97, 155
Video On Demand (VOD), 43, 49-50, 263
Video Processing, 2, 280, 363
Video Scaling, 278
Video Surveillance, 32
Video Warping, 176, 280
Viewer Profile, 40
Virtual Reality Modeling Language (VRML), 87, 174
Warping, 176, 280
Web Browser, 283, 340-341
Web Interface, 287, 289
Widescreen, 62, 293
Windows Media (WM), 95, 97, 125, 129
Windows Media Audio (WMA), 95, 125, 129
Wireless Keyboard, 288, 298
Workflow, 194-195, 202
Workflow Automation, 202
Workflow Management, 194-195

# Althos Publishing Book List

| Product ID | Title | # Pages | ISBN | Price | Copyright |
|---|---|---|---|---|---|
| **Billing** | | | | | |
| BK7781338 | Billing Dictionary | 644 | 1932813381 | $39.99 | 2006 |
| BK7781339 | Creating RFPs for Billing Systems | 94 | 193281339X | $19.99 | 2007 |
| BK7781373 | Introduction to IPTV Billing | 60 | 193281373X | $14.99 | 2006 |
| BK7781384 | Introduction to Telecom Billing, 2nd Edition | 68 | 1932813845 | $19.99 | 2007 |
| BK7781343 | Introduction to Utility Billing | 92 | 1932813438 | $19.99 | 2007 |
| BK7769438 | Introduction to Wireless Billing | 44 | 097469438X | $14.99 | 2004 |
| **IP Telephony** | | | | | |
| BK7781311 | Creating RFPs for IP Telephony Communication Systems | 86 | 193281311X | $19.99 | 2004 |
| BK7780530 | Internet Telephone Basics | 224 | 0972805303 | $29.99 | 2003 |
| BK7727877 | Introduction to IP Telephony, 2nd Edition | 112 | 0974278777 | $19.99 | 2006 |
| BK7780538 | Introduction to SIP IP Telephony Systems | 144 | 0972805389 | $14.99 | 2003 |
| BK7769430 | Introduction to SS7 and IP | 56 | 0974694304 | $12.99 | 2004 |
| BK7781309 | IP Telephony Basics | 324 | 1932813098 | $34.99 | 2004 |
| BK7781361 | Tehrani's IP Telephony Dictionary, 2nd Edition | 628 | 1932813616 | $39.99 | 2005 |
| BK7780532 | Voice over Data Networks for Managers | 348 | 097280532X | $49.99 | 2003 |
| **IP Television** | | | | | |
| BK7781362 | Creating RFPs for IP Television Systems | 86 | 1932813624 | $19.99 | 2007 |
| BK7781355 | Introduction to Data Multicasting | 68 | 1932813551 | $19.99 | 2006 |
| BK7781340 | Introduction to Digital Rights Management (DRM) | 84 | 1932813403 | $19.99 | 2006 |
| BK7781351 | Introduction to IP Audio | 64 | 1932813519 | $19.99 | 2006 |
| BK7781335 | Introduction to IP Television | 104 | 1932813357 | $19.99 | 2006 |
| BK7781341 | Introduction to IP Video | 88 | 1932813411 | $19.99 | 2006 |
| BK7781352 | Introduction to Mobile Video | 68 | 1932813527 | $19.99 | 2006 |
| BK7781353 | Introduction to MPEG | 72 | 1932813535 | $19.99 | 2006 |
| BK7781342 | Introduction to Premises Distribution Networks (PDN) | 68 | 193281342X | $19.99 | 2006 |
| BK7781357 | IP Television Directory | 154 | 1932813578 | $89.99 | 2007 |
| BK7781356 | IPTV Basics | 308 | 193281356X | $39.99 | 2006 |
| BK7781389 | IPTV Business Opportunities | 232 | 1932813896 | $24.99 | 2007 |
| BK7781334 | IPTV Dictionary | 652 | 1932813349 | $39.99 | 2006 |
| **Legal and Regulatory** | | | | | |
| BK7781378 | Not so Patently Obvious | 224 | 1932813780 | $39.99 | 2006 |
| BK7780533 | Patent or Perish | 220 | 0972805338 | $39.95 | 2003 |
| BK7769433 | Practical Patent Strategies Used by Successful Companies | 48 | 0974694339 | $14.99 | 2003 |
| BK7781332 | Strategic Patent Planning for Software Companies | 58 | 1932813322 | $14.99 | 2004 |
| **Telecom** | | | | | |
| BK7781313 | ATM Basics | 156 | 1932813136 | $29.99 | 2004 |
| BK7781345 | Introduction to Digital Subscriber Line (DSL) | 72 | 1932813454 | $14.99 | 2005 |
| BK7727872 | Introduction to Private Telephone Systems 2nd Edition | 86 | 0974278726 | $14.99 | 2005 |
| BK7727876 | Introduction to Public Switched Telephone 2nd Edition | 54 | 0974278769 | $14.99 | 2005 |
| BK7781302 | Introduction to SS7 | 138 | 1932813020 | $19.99 | 2004 |
| BK7781315 | Introduction to Switching Systems | 92 | 1932813152 | $19.99 | 2007 |
| BK7781314 | Introduction to Telecom Signaling | 88 | 1932813144 | $19.99 | 2007 |
| BK7727870 | Introduction to Transmission Systems | 52 | 097427870X | $14.99 | 2004 |
| BK7780537 | SS7 Basics, 3rd Edition | 276 | 0972805370 | $34.99 | 2003 |
| BK7780535 | Telecom Basics, 3rd Edition | 354 | 0972805354 | $29.99 | 2003 |
| BK7781316 | Telecom Dictionary | 744 | 1932813160 | $39.99 | 2006 |
| BK7780539 | Telecom Systems | 384 | 0972805397 | $39.99 | 2006 |

| Product ID | Title | # Pages | ISBN | Price | Copyright |
|---|---|---|---|---|---|
| **Wireless** | | | | | |
| BK7769434 | Introduction to 802.11 Wireless LAN (WLAN) | 62 | 0974694347 | $14.99 | 2004 |
| BK7781374 | Introduction to 802.16 WiMax | 116 | 1932813748 | $19.99 | 2006 |
| BK7781307 | Introduction to Analog Cellular | 84 | 1932813071 | $19.99 | 2006 |
| BK7769435 | Introduction to Bluetooth | 60 | 0974694355 | $14.99 | 2004 |
| BK7781305 | Introduction to Code Division Multiple Access (CDMA) | 100 | 1932813055 | $14.99 | 2004 |
| BK7781308 | Introduction to EVDO | 84 | 193281308X | $14.99 | 2004 |
| BK7781306 | Introduction to GPRS and EDGE | 98 | 1932813063 | $14.99 | 2004 |
| BK7781370 | Introduction to Global Positioning System (GPS) | 92 | 1932813705 | $19.99 | 2007 |
| BK7781304 | Introduction to GSM | 110 | 1932813047 | $14.99 | 2004 |
| BK7781391 | Introduction to HSPDA | 88 | 1932813918 | $19.99 | 2007 |
| BK7781390 | Introduction to IP Multimedia Subsystem (IMS) | 116 | 193281390X | $19.99 | 2006 |
| BK7769439 | Introduction to Mobile Data | 62 | 0974694398 | $14.99 | 2005 |
| BK7769432 | Introduction to Mobile Telephone Systems | 48 | 0974694320 | $10.99 | 2003 |
| BK7769437 | Introduction to Paging Systems | 42 | 0974694371 | $14.99 | 2004 |
| BK7769436 | Introduction to Private Land Mobile Radio | 52 | 0974694363 | $14.99 | 2004 |
| BK7727878 | Introduction to Satellite Systems | 72 | 0974278785 | $14.99 | 2005 |
| BK7781312 | Introduction to WCDMA | 112 | 1932813128 | $14.99 | 2004 |
| BK7727879 | Introduction to Wireless Systems, 2nd Edition | 76 | 0974278793 | $19.99 | 2006 |
| BK7781337 | Mobile Systems | 468 | 1932813373 | $39.99 | 2007 |
| BK7769431 | Wireless Dictionary | 670 | 0974694312 | $39.99 | 2005 |
| BK7780534 | Wireless Systems | 536 | 0972805346 | $34.99 | 2004 |
| BK7781303 | Wireless Technology Basics | 50 | 1932813039 | $12.99 | 2004 |
| **Optical** | | | | | |
| BK7781386 | Fiber Optic Basics | 316 | 1932813861 | $34.99 | 2006 |
| BK7781329 | Introduction to Optical Communication | 132 | 1932813292 | $14.99 | 2006 |
| BK7781365 | Optical Dictionary | 712 | 1932813659 | $39.99 | 2007 |
| **Marketing** | | | | | |
| BK7781318 | Introduction to eMail Marketing | 88 | 1932813187 | $19.99 | 2007 |
| BK7781322 | Introduction to Internet AdWord Marketing | 92 | 1932813225 | $19.99 | 2007 |
| BK7781320 | Introduction to Internet Affiliate Marketing | 88 | 1932813209 | $19.99 | 2007 |
| BK7781317 | Introduction to Internet Marketing | 104 | 1932813292 | $19.99 | 2006 |
| BK7781317 | Introduction to Search Engine Optimization (SEO) | 84 | 1932813179 | $19.99 | 2007 |
| BK7781323 | Web Marketing Dictionary | 688 | 1932813233 | $39.99 | 2007 |
| **Programming** | | | | | |
| BK7781300 | Introduction to xHTML: | 58 | 1932813004 | $14.99 | 2004 |
| BK7727875 | Wireless Markup Language (WML) | 287 | 0974278750 | $34.99 | 2003 |
| **Datacom** | | | | | |
| BK7781331 | Datacom Basics | 324 | 1932813314 | $39.99 | 2007 |
| BK7781355 | Introduction to Data Multicasting | 104 | 1932813551 | $19.99 | |
| BK7727873 | Introduction to Data Networks, 2nd Edition | 64 | 0974278734 | $19.99 | 2006 |
| **Cable Television** | | | | | |
| BK7780536 | Introduction to Cable Television, 2nd Edition | 96 | 0972805362 | $19.99 | 2006 |
| BK7781380 | Introduction to DOCSIS | 104 | 1932813802 | $19.99 | 2007 |
| BK7781371 | Cable Television Dictionary | 628 | 1932813713 | $39.99 | 2007 |
| **Business** | | | | | |
| BK7781368 | Career Coach | 92 | 1932813683 | $14.99 | 2006 |
| BK7781359 | How to Get Private Business Loans | 56 | 1932813594 | $14.99 | 2005 |
| BK7781369 | Sales Representative Agreements | 96 | 1932813691 | $19.99 | 2007 |
| BK7781364 | Efficient Selling | 156 | 1932813640 | $24.99 | 2007 |

# Order Form

Phone: 1 919-557-2260
**Fax: 1 919-557-2261**
404 Wake Chapel Rd., Fuquay-Varina, NC 27526 USA

Date:_____

Name:_____ Title:_____
Company:_____
**Shipping Address:**_____
City:_____ State:_____ Postal/ Zip:_____
**Billing Address:**_____
City:_____ State:_____ Postal/ Zip _____
Telephone:_____ Fax:_____
Email:_____
**Payment (select):** VISA ___ AMEX ___ MC ___ Check ____
Credit Card #: _____Expiration Date: _____
Exact Name on Card: _____

| Qty. | Product ID | ISBN | Title | Price Ea | Total |
|------|-----------|------|-------|----------|-------|
|      |           |      |       |          |       |
|      |           |      |       |          |       |
|      |           |      |       |          |       |
|      |           |      |       |          |       |
|      |           |      |       |          |       |
|      |           |      |       |          |       |
|      |           |      |       |          |       |
|      |           |      |       |          |       |
|      |           |      |       |          |       |
|      |           |      |       |          |       |

| | |
|---|---|
| Book Total: | |
| Sales Tax  (North Carolina Residents please add 7% sales tax) | |
| Shipping: $5 per book in the USA, $10 per book outside USA (most countries).  Lower shipping and rates may be available online. | |
| **Total order:** | |

**For a complete list please visit**
**www.AlthosBooks.com**

Printed in the United Kingdom
by Lightning Source UK Ltd.
129806UK00002B/55/A